ENVIRONMENT IN CRISIS

Selected Essays on Somali Environment

Ahmed Ibrahim Awale

Copyright©2016, Ahmed Ibrahim Awale

All rights Reserved

ISBN: 978-87-995208-5-5

Second Edition

All rights reserved. No part of this book may be reproduced, in any form or any means, without permission in writing from the author.

The author can be can be contacted through this e-mail:
aiawaleh@gmail.com

Contents

ACKNOWLEDGEMENT 5
DEDICATION 7
FOREWORD 9
PREFACE 11
The Barren Tree 14
Zizyphus Tree 17
Environmental Contamination of Ayaha Valley 21
Surad Mountain Forest in Somaliland: A Cry for Help 26
The Somali Wild Ass 30
The Vanishing Trees of Hargeysa 33
Denudation of Batalaale Conocarpus lancifolius Plantation
.... 41
Mesquite (Prosopis juliflora) 44
Deforestation of the Mountains of Sheikh 49
Charcoal Production – Great Threat to Somali Livelihoods
.... 53
Climate Change 56
Al Wabra Wildlife Preservation (AWWP) 62
High Mortality of Acacia Trees in the Coastal Areas 64
He who Plants a Tree, Plants Hope 67
Declining Mist Levels over Gacan Libaax 70
Vanishing Bees 85
The death of the Famous Dheen Tree 88
Toxins from Tanneries 92
Ye-eb (Cordeauxia edulis) 96
The Pillage and Plunder of Somali Marine Resources 99
The Link between Droughts and Deforestation 103

Honey Adulteration and Testing its Purity	107
Me and My Toothbrush Tree	110
Rapshie Island	115
Essay on Environment	119
Where Have the Birds Gone?	123
Date Palm (Phoenix Dactylifera)	130
Domestication of Henna in Somaliland	137
Miscellaneous Environmental Reflections	143
Explanatory Descriptions of the Plates	170
List of Common Plants	173
Foreign Trees	181
Edible Wild Food Plants	182
Names of some fishes in Somali waters	186
Diseases of Livestock and burden animals	187
Selected list of common mammals	189
Glossary	191
Bibliography	201
Index	206
Other books by the author	210

ACKNOWLEDGEMENT

I would like first to thank the Almighty Allah, for enabling me to accomplish this task.

I would also like to express my appreciation to:

Candlelight for Health, Education and Environment, a prominent environmental organization based in Somaliland, for giving me the opportunity, through my employment, to delve into the field of environmental studies and, at the same time, put it into practice. Without such support and encouragement from my work colleagues, completing this book would not have been a simple matter.

The members of the natural resource management of CLHE (Abdirazak Bashir, Ahmed Jama Sugulle, Asha Abdirahman and Abdiqani Suleiman) – my companions in the field of ecology whom we shared valuable experiences and together carried out many practical interventions. Special thanks go to Ms. Shukri Haji Ismail, the former Chairperson of Candlelight, for her encouragement and environmental stewardship.

Many other friends also provided their support and inputs in their different areas of expertise. They include Jama Musse Jama (for shouldering the task of publishing the first edition), Ahmed Gaheir Jama (for his encouragement and writing a fitting foreword), Aden Yusuf Abokor, Ingrid Hartmann, Eric Schewennesen, and Chris (Roble) Print for proofreading, and commenting on the various sections of the book.

Marwan El Azzouni and Giuseppe Orlando for allowing me to include their article "Surud Mountain: A Cry for Help" in this book, and for translating it into Somali; Abdi Ali Jama for his beautiful account on Buurta Rabshiga (Maydh Island); Osman Mohamed Ali for writing an article on date palm cultivation in Puntland, Dr. Esse Nur Liban for enriching the list of livestock diseases; and Mohamed Abdullahi Jama's contribution to the section on Wild Food Plants in the book.

Finally, Nimo Abdillahi Hassan for translating the first essay of the book, *the Barren Tree,* from the original Somali, and Hamda Mohamed Hussein Egal's for assisting in typing some sections of the book.

Ahmed Ibrahim Awale

DEDICATION

I dedicate this book to the generations yet to be born who will need to live in a protected and sustainable environment, and a life that is no less fulfilling (if not better) than the one we live in today.

FOREWORD

It has been a while that I personally felt loneliness and helplessness in seeing the on-going onslaught on our environment. I have been teaching a course on environment and agriculture at Amoud University for three academic years now and I still do, however, my naivety and ingenuousness never let me look beyond the horizon. Nonetheless, I came to realize that I should not entertain lonesome dreams about the situation of our environment, as I, finally, came to know that I have a companion - someone who will leave an indelible imprint on the environmental landscape in the region.

I am proud of Ahmed Ibrahim Awale for such a benevolent and pioneering dedication of this book, which is probably the single most important task facing every one of us on daily basis regarding our deteriorating environment. Again, Ahmed, I must applaud you for such a good work.

In short, Ahmed, it is a difference-making effort that hopefully will guide many others like you in this direction. I personally envisage that this work will contribute to better environmental awareness and improved stewardship towards our ailing environment. The Chinese ideogram for "crisis" combines the character *wei*, meaning *danger*, with character *ji* denoting *opportunity*. Therefore, WEI JI is a good description of our global environment situation. We are dangerously close pushing biological communities and biophysical processes beyond the point from which they can recover but, at the same time, there is always *the opportunity* to "fight" back and resist the danger. This Chinese ideogram has contributed to the creation of notable environmental awareness. So, Ahmed, yours will create similar awareness among our communities.

Your worthwhile contribution and the material collection are a valuable asset and it is highly way beyond any words of any kind. I should humbly say to you, may Allah (SWT) reward you for your great effort.

Prof. Ahmed Gaheir Farah
Lecturer, Amoud University

PREFACE

As far as we know, the Earth is the only planet that nurtures the existence of life. The launching of Sputnik 1 in 1957, the world's first earth-orbiting satellite, opened the door for the "space race", chiefly between the Soviet Union and the United States. Since then, scientists and astronomers worldwide have spent a lot of time and resources in search of, among other things, an existence of life other than the one on our planet. However, the stark truth, so far, is that there is no trace of any form of life known to exist outside our planet Earth.

Realizing this fact, that we (inhabitants of the Earth) are alone and lonely in a vast "non-living" cosmos, reminds us the finiteness of the Earth (as ourselves) and how vulnerable it is. We are living in a world of ever-diminishing resources, expanding needs and deteriorating quality of life.

The earth can be compared to a spaceship with its crew. By analogy, the resources (food, water and air) in it are limited, finite and subject to pollution while the crew is growing larger day after day.

The responsibility of protecting this fragile environment on earth lies on the shoulders of humankind. In other words, we are God's viceroy on earth and stewards of creation. Therefore, we have to ensure that the ecosystems and biophysical processes function in harmony to maintain the equilibrium and the health of the planet and of all living organisms. Yet our actions are leading to destruction rather than sustaining life and ensuring its continuity for meeting the needs of the present, without compromising the needs of the future generations.

In writing this book, I gave more attention to the status of Somali environment while, at the same time, attempted to present to the reader some of the present-day and future global environmental challenges facing our planet, climate change in particular. I wrote the articles published in this book in different period. These articles were selected from dozens of environmental essays which appeared in *Deegaankeenna* Newsletter, published by Candlelight for Health, Education and Environment – a non-governmental organization based in Hargeysa, Somaliland. The approach I followed in selecting the following articles was to ensure that notable environmental challenges – both local and global – are reflected in the book.

Although I tried to capture and sum up in this book the different aspects of our environment, the discerned reader may notice that the author is biased towards terrestrial environment, and the mention of marine resources is very limited. This does not mean that the sea is less important to people than land. The basis of my choice, however, was the fact that the most ideal habitat of humankind is the Land and, therefore, their actions on land could influence or determine the way they utilize or exploit the sea. If we choose to protect and conserve our land-based natural resources and life forms, undoubtedly we will also take better care of our marine resources.

My objective in writing this book is to put across to the reader, in particular, and to the whole community in general, some glimpses on the status of our environment - its beauty and diversity and capacity to sustain itself and support livelihoods, and its vulnerabilities and struggle against unprecedented assaults. This is a *cry for help* translating the plight of our environment - a message whispering within our conscience, into the hope it may trigger concerted actions towards the mitigation of the on-going onslaught on the natural resources of the country.

I have also taken the mammoth task of rendering this book into Somali, in the hope that the messages they contain may reach far and wide and audiences from different walks of life and backgrounds such as students, researchers, decision-makers and the community at large.

Enjoy your reading!

Ahmed Ibrahim Awale
Hargeysa, 15th January, 2016

1

The Barren Tree

What happened to the tree that "thought" it was saved from barrenness (childlessness)!

In 1988, somewhere in downtown Hargeysa, there stood an old *Maraa* (*Acacia nilotica*) tree. Whenever I walked by, I used to remember a verse from the Qur'an based on prophet Zachariah's prayers *'O my Lord! Leave me not without offspring; Thou art the best of inheritors." (Quran: 21:89)*

This was because, all the falling seeds, which would have germinated and then might have replaced this old tree, were always trodden under the feet of many shoppers and vehicle tires. Then the civil war came (1988-1990), when everybody in the city fled for his life to safety; thus emptying the city. The tree perished in the heavy shelling that fell on the city. However, three new saplings sprouted from the seeds shed by the old tree. These trees found an environment much conducive for its growth, as there were neither animal to feed on them nor human feet to trample on them. They grew wholesomely, full with promise and optimism, and hoping to provide immense benefits to humans; or so they thought!

The leaves of the tree, its pods and shoots are important source of fodder. The pods, which are locally known as "gaydhe", ripen during the dry season, when fodder material is scarce. Pastoralists shake its branches to fell the pods and these are relished by goats and camels. It is fast growing compared to many other acacias. It is also used as a pioneer species in land rehabilitation and as a barrier to desertification. Birds build their nests amid its branches.

Then the civil war came to end, and people started returning to Hargeysa. Immediately, they occupied every open space available – houses, markets and streets alike. These trees grew in front of some shops and stores. Immediately *Qat*[1] sellers and girls, who migrated from the intricate pastoral life, set up market stalls under the trees. Unfortunately, after a short while, arguments arose among the stall owners operating under the three young trees and the shopkeepers. They exchanged heated words and quarrelled with arguments in the following form:

A shopkeeper: 'You are causing a hindrance to my business. Move your stall away from the front-face (*wajahad*) of my shop'.

A stall owner: 'It is not you who owns - this tree and the street!'

Thus, it came to pass that the trees became the prime target, because they were thought to be the bone of contention. The shopkeepers reasoned: If the trees were not there, the areas could have been conflict free; or so they thought!

Then one night, one of the shop owners decided to cut one of the trees and then actually did it. The following night, another shopkeeper did the same thing and cut the second

---

[1] Qat *(Catha edulis)*: An evergreen shrub and a mild stimulant whose leaves are chewed that cause euphoric effects.

tree. Finally, the last tree disappeared a few nights later, most probably "slaughtered" by the third shopkeeper.

Consequently, the old tree that 'thought' to have been saved from barrenness could not escape from that same fate.

2

Zizyphus Tree

> "There is a big Zizyphus tree on the surface of the moon with leaves matching, at any given point of time, the number of living people on the planet. When a baby is born, the tree brings forth a new leaf and then when a person dies, his/her leaf withers away and falls from the tree"._____ *a popular Somali folklore.*

Zizyphus mauritiana (locally known as *Gob* and internationally as Chinese date, Jujube, Indian Plum, and Indian Cherry) is a tree very common to the whole Sahelian zone, the Horn of Africa and in all the drier parts of Asia. In the Somali country, it can be found in all ecological zones, but more commonly in the drier areas, particularly on the banks of seasonal watercourses. The evergreen drought resistant tree has important ecological and socio-economic benefits. The tree produces a delicious and nutritious fruit (*Midhaha Gobka*) that is a good source of vitamin C.[2] Refreshing drink can also be made from the fruits. The leaves are palatable and are liked very much by livestock, particularly goats and camels. In India, the leaves are used to feed the silkworms reared for silk production. The *Gob* tree produces good firewood, charcoal and versatile hardwood. It is an ideal tree from land reclamation and good for use as a life fence. It is believed

---

[2] The unripe green fruit is locally known as *"ugur"*, when they turn to yellow they are called *"caddays"*, and when they ripen and turn to red they are termed as *"hoobaan"*.

that honey produced from Zizyphus plants is of a high quality that fetches a good price in the Gulf States, compared to other honeys that bees source from plants other than Zizyphus. *Qasil*– a popular natural shampoo and a pale green body and face mask – can be made from dried and then powdered leaves. *Qasil* is also reputed to nourish, beautify hair and combat dandruff. In the Islamic traditional medicine, the leaves are used in exorcising evil spirits. Recent medical research suggests that *Gob* maybe effective in inhibiting skin cancer by inducing cell death in melanoma. It is also believed to retard greying of the hair. Muslims use *Qasil* for cleaning dead bodies in preparation for burial as it is believed that it delays the decomposition process. It also served as anti-lice pesticide, used by women by fumigating through the smoke of its dry leaves and firewood.[3] *Gob* trees live and remain productive for several centuries.

Zizyphus mauritiana belongs to the family *Rhamnaceae*, of which (it is supposed) "Christ's Crown of Thorns"[4] was made. There are two more species of Zizyphus found in the Somali lands namely, *Zizyphus hamur* (Somali: *Xamudh*) that is found in the coastal areas and *Zizyphus spina christi* (*Gob-yar*).

A different meaning for the word "*Gob*" is 'noble'. One can say someone is "*Gob*" and his /her deed is "*gobanimo*". Similarly, the independence/freedom of a country from a foreign power is "*gobanimo*" while the opposite of "*gob*" is

---

[3] Similarly, the resin from Xabaghedi (*Commiphora guidotti*) was burned by women as an incense to give a pleasing smell to their intimate parts.

[4] As per the Christian scriptures, this is the thorny wreath plaited by the soldiers of Pilate and put on the head of Jesus, when they mocked Him as king of the Jews (Matthew 27; Mark 15; John 19).

"*gun*" which literally means "root" – an allusion to baseness or meanness. Therefore, one can say some one is "*gun*" and his/her deeds "*gunnimo*". In a political context, "*gunnimo*" means "the state of being under colonization" while in the Somali culture it means "the state of being servile". If we put the above in the context of care for the *Gob* tree, to cut a *Gob* (noble) tree is to make it "*gun*" (mean, servile).

Therefore, cutting a Gob tree is culturally an unacceptable action.

The *Gob* plant is mentioned in the Holy Qur'an as a symbolic heavenly bliss[5] and sometimes called "Tree of Life": A Somali popular folklore says that there is a big *Gob* tree on the moon with leaves matching, at any point of time, the number of living people on the planet. It is believed that when a baby is born, the tree brings forth a new leaf and when person dies, *his/her* leaf withers away and falls from the tree. That is the reason why people warn an arrogant and troublesome person who is at the verge of killing someone or being killed: "*Won't you stop this! Or is it that your Leaf (on the moon) is about to wither away and die!*"

Here is an interesting story indicating how Somalis value the plant: There were two agro-pastoral households settled near Oodweyne town. One of the families was very poor by local standards, but had one Gob tree in his possession that was known for its good fruits and other socio-economic benefits. The family had a son who fell in love with a girl from the other household. Then the family of

---

[5] Holy Qur'aan, Al-Najam: 53-13-18

the groom proposed to offer the Gob tree as a pride price to the parents of the girl, which, the later then accepted with great delight. The land on which the tree stood also went to the family of the bride, but still the *Gob* tree was the 'jewel in the crown'.

As we are in an era of renewed interest in the use of nature-based products, it is worthwhile that the multi-faceted uses of Gob – be it medicinal, cosmetic, spiritual and/or socio-economic – be appreciated. Hence, the necessity for its protection, propagation and wider dissemination.

3

Environmental Contamination of Ayaha Valley of Hargeysa, Somaliland

> "Ayax teg, eel se reeb" {Locusts are gone, but left destruction and future threat behind its path} *A Somali proverb*

Magaalo Qallooc is located at the south-western corner of Hargeysa and has been the Desert Locust Control Organization for East Africa (DLCO-EA) compound since the British colonial days. This compound housed toxic chemicals which was used, when need arises, for areal spray. Since it is establishment, it used to be a highly guarded site with round the clock security personnel.

Since the return of the war-displaced people, as well as many others to Somaliland in the early 1990s, a large section of the community of Hargeysa has been expressing their concern about the spillage of toxic chemicals in the ex-DLCO-EA centre. Situated five km downhill and southwest of Hargeysa, the site was extensively bombed and destroyed during the 1988-1990 war. The centre was a storage site for various chemicals used to control migratory locusts in the region. It is estimated that more than 80,000 litres of chemicals suspected to be pesticides were poured into the ground and the drums used by the area residents for water storage and other household uses.

In August 2003, two consultants from Kenya Plant Health Inspectorate Services (KEPHIS) conducted a study on the environmental contamination of Ayaha Valley and then presented the findings to a number of concerned agencies and institutions in Somaliland. The consultancy was requested by the Municipality of Hargeysa and was funded by the United Nations Development Programme (UNDP). The study area covered the Ayaha Valley, where DLCO-EA had its facilities, all the way downstream to the Maroodi Jeex seasonal water course, reaching the Irish crossing at Haraf Restaurant.

The objectives of the environmental consultancy in Ayaha Valley were:

- To assess the area and level of pesticide contamination in the valley surrounding the ex-DLCO compound;
- To sample and analyze soil and water samples from the contaminated area;
- To provide a comprehensive report on the methodology, results and interpretation and recommendations regarding suitability of the area for settlement of returnee persons and possible future remedial actions.

The report shows that the pesticides that have been spilled in the site are highly toxic and are having health and environmental risks. These include Dieldrin, Alpha-HCH, Beta-HCH, Eldrin, Beta-Endosulfan, DDT, Heptachlor and Lindane – most of them banned worldwide in the 1980's.

Soil samples for laboratory examination were collected from 19 sites between the ex-DLCO-EA compound and Bar Haraf, along the course of the small seasonal riverbed draining the Ayaha valley.

According to the report, the high concentration of organochlorine pesticides detected in the soil samples indicates high contamination of the entire Ayaha Valley and classified it as an example of catastrophy of mass proportion. The organochlorine pesticides are persistent chemicals which can last in the environment for a long time and can cause acute and chronic diseases which can damage the nervous system in human beings. Some may eventually cause cancer.

According to the World Health Organization (WHO) in Somaliland, deformities in maternity cases and high abortion rate have been experienced in this area. This could be the result of the high chemical contamination in and around the ex-DLCO compound.

During rainy seasons, the chemicals can be washed away downstream into Hargeysa proper. This poses a big danger to the city residents and those living around the valley. However, the river itself may get flooding from rains outside the city and allow the chemicals to be transported further downstream. There is a strong smell of pesticides that can be felt at 500m away from the contaminated compound. This again poses a great danger to the nearby Ayaha Primary School and people living in the wind direction. Prolonged exposure to the smell can be injurious to human beings.

The report concludes with the following recommendations:

<u>Short-term</u>

- The highly contaminated ex-DLCO compound should be fenced immediately to prevent entry by both human beings and livestock.
- The fenced compound should have only one gate manned by a full time watchman or guard;
- The people currently living around the contaminated compound and the valley below (downstream) should be moved and relocated to a safe place. They should be educated and fully informed of the danger of chemical contamination and poisoning. All human beings and

their livestock should be discouraged from settling or moving about in the heavily contaminated area;
- Roofing of all the contaminated area will prevent rainfall washing away the chemicals downstream.

- <u>Long-term</u>
- Proper cleaning of the contaminated area is essential. The cleaning can be done by digging and removing the whole contaminated soil and dumping it in a desert away from human settlement or incinerate it in very high temperature.
- An international conference or meeting of all stakeholders should be convened by either FAO or UNDP as early as possible to discuss and deliberate on the whole environmental contamination of Ayaha Valley and agree on possible long term remedial actions.
- In order to ascertain the health of the residents residing around the valley, WHO, in collaboration with FAO and UNDP, should establish a joint monitoring and analysis of blood and /or breast milk from the residents in the area. Also due to the food chain phenomenon, milk and meat from goats and other domesticated animals kept around the compound should be analyzed for trace and levels of chemical residues.

{The article was adapted from a report "Environmental Contamination of Ayaha Valley, Hargeysa, Somaliland", Dr. Rhonest Ntayia & Mr. James Kinyua, KEPHIS/UNDP Project, July 2003}

Postscript:

Years after this report was released, it is worth asking: What steps have been taken in addressing the problem of chemical contamination in Ayaha Valley? And what has been done to implement the recommendations of the experts?

A masonry fence wall was built around the site; unfortunately the surveillance and guarding is weak. The wall is broken in some areas and one can sometimes see goats grazing inside the compound and children jumping over the wall. Before the war, the whole area surrounding the site was a 'no-go' zone; however, till the time of writing this article human settlement has been encroaching on the site from all directions. The deadly chemicals are still lying inside the compound.

Unfortunately, neither the concerned government institutions nor development agencies seem to be interested in clearing this sinister mess and relieving the residents from the ever-present fear and worry caused by the presence of the chemicals in the area. On the other hand, the appreciation of land prices which resulted from the increasing population of Hargeysa makes those residents ignore any attempt to relocate them to safer places, unless concerted efforts are made to aware them of the inherent risks as well as motivational packages such as new land and access to basic social services.

4

Surad Mountain Forest in Somaliland: A Cry for Help
By: *Marwan El Azzouni and Giuseppe Orlando (January, 2004)*

There was something strange about the fire that our guide made while preparing tea in our camp 5 km north of Ceerigaabo. While sipping our sweetened tea, the smell kept lingering in the air; it was sweet, as sweet as innocence. It was on a slightly chilly evening last November while we were waiting to enter the Surad Mountains in Somaliland, one of the last seemingly untouched places on this planet. We had actually left behind all (our) fears in Cairo as we set off to Somaliland, following the footsteps of early explorers. We had decided to make the visit in search of a few elusive stapeliad species and also to view the majestic *Aloe eminens* in its habitat. The Surad Mountains are in the northern part of Somaliland, in the Sanaag region. The highest peak is Mount Shimbiris, at 2416 m the highest mountain in all of Somalia. Early the next morning, after a sleepless night in anticipation of all the wonders we would eventually see at daybreak, we headed towards the mountain range; then reaching the top of a plateau that plunges downwards towards the coast. We passed fields of *Aloe scobinifolia* and *Euphorbia ballyi* before entering an area of very lush vegetation at the mouth of the famous *Tabah (Tabca)* Gorge. We could see the gorge plunge deep into the forest to our right. Had we the choice, we would have left the car behind and gone deep down exploring that fabulous gorge, where huge *Dracaena schizantha* trees hang on the cliffs.

These mountains are part of the Somali Montane Xeric Woodland ecoregion that stretches along the northern coast of the Horn of Africa from the Shimbiris to Raas Caseyr, continuing some 300 km south along the Somali coastal plain. Although part of the Somali-Masai regional centre of endemism, this area also contains remnant plant species linking it to Mediterranean, Micronesian and Afromontane regions. A special environment is created by the mountain chain facing north, accumulating plenty of mist from the sea. The Daloh-Shimbiris area receives the highest rainfall in Somaliland, over 700 mm each year, favouring the evolution of a unique and extremely diverse flora.

Entering the Tabah (*Tabca*) gorge was like reaching paradise on earth. We stopped the engine and both jumped out of the vehicle, each into a different direction. As we disappeared into the vegetation, we started to see and hear things we were not aware of when the engine was running. Colours, smells, sounds: it was truly alive. Huge *Juniperus procera* trees more than 20 m high tower among lush mixed woodland, whose fresh green is heavily spotted with the grey crowns of *Dracaena schizantha* and the striking red flowers of *Aloe eminens (see photo no. 5)*.

We saw scattered populations of flowering *Aloe albovestita* and patches of *A. hildebrandtii*. Plants (also) seemed to sprout from under every rock and inside every crack. (We saw) bushes of *Buxus hildebrandtii*, several species of Commiphora, and succulents - *Kalanchoe spp., Senecio spp.* and probably a new species of *Huernia*. It was a botanical heaven.

Unfortunately, since a rumour (had) spread as wildfire in the area that we were diamond hunters, we could not risk getting off the main track that would eventually snake itself through the mountain range. In only 10 km as the crow flies, we passed from the juniper forests of the misty high altitudes crossing very distinct zones of vegetation, down to the extremely arid plains, *Guban* (meaning burnt in English), ending up in Maydh on the coast. The track, however, winds down the mountain range for about 40-50 km. The landscape and vegetation are amazing - spectacular woodlands of frankincense (*Boswellia frereana*), often growing on huge boulders or in vertical cliffs, occasional *Pyrenacantha malvifolia* with fat caudices that can reach over one metre in diameter, *Commiphora spp.* with their strong smell and blue, white or grey bark, Aloes, *Euphorbias* and several other strange xerophytic species. Looking for small stapeliads entails bending and looking under shrubs, rocks and in cracks for these shy plants - this was the origin of the rumour.

We had tried to explain to the surprised locals that we were looking for *Ubah* (flower, in the Somali language), but no one believed us.

For the first few days in the range, one is taken by the beauty and variety of flora, a natural botanical garden. It was on the fourth day, as we were sitting on a ledge overlooking the great cliffs that surround the Tabah (*Tabca*) Gorge, that it hit us. *Thud ... thud . thud.* Yes, it was an axe chopping a tree. We turned our ears and eyes to locate the source of this logging. Far away in a distance we saw a plume of smoke rising from the thick canopy. From then on and for the next four days, we saw only destruction, juniper trees felled like matchsticks, huge 100-year old *Commiphoras* cut for charcoal and building material, areas cleared for Qat plantations, total destruction of a pristine and extremely exotic forest. It was sad to discover that the origin of the sweet smell we enjoyed a few days before in our camp fire was not so innocent: aromatic *Commiphoras* turned into charcoal. Old trunks of *Juniperus procera* cut to make firewood, charcoal or building materials were not an uncommon sight.

Somaliland is short of cash, particularly after the Gulf States set a ban on livestock exports. Somali authorities have recently given licensing permits to a few greedy companies to burn the forests for export-oriented charcoal production. According to recent reports, the areas most affected are those (to the) east of Hargeysa (the capital), most probably around the Golis Mountain range that is closer to Berbera, the main seaport. The destruction in the Surad Mountains is more local in scale, and can be avoided through careful education of the local tribes that live there and increasing their awareness regarding their natural wealth. However, if the charcoal production companies eventually reach the Surad Mountains, an area of immense beauty and extreme natural diversity will be lost forever.

Daloh forest is the best preserved *Juniperus procera-Dracaena schizantha* mixed woodland in the Horn of Africa and it well deserves to be declared a World Heritage Natural Site. This article is a cry for help for anyone interested in saving such a unique and exotic place rich in plant and animal endemics from disappearing before having the opportunity to be even studied. We are planning to arrange for a scientific expedition in the near future, in cooperation with local authorities, as most of the mountain range remains yet to be explored.

5

The Somali Wild Ass

> "The Wild Ass which grazed on the lush grass, arrogantly stroke the elephant with its hind leg"_____Haji Aadan Af-qallooc

In this article, I will attempt to discuss the status of the endemic Somali Wild Ass (*Equus africanus somaliensis*), which has been alerted by the World Conservation Union (IUCN) as critically endangered animal known to have inhabited in the Somali territory, Djibouti and Eritrea. The Somali Wild Ass (see plate # 21) is one of three sub-species (types) of African wild ass. It is the ancestor of the domestic donkey, and has a similar stocky body shape. Overall, the species is the smallest of the wild equids (horses, assess, and zebras). The Somali Wild Ass stands about 1.20m at the shoulder and weighs about 270kg. It is mostly grey in colour, with a white belly. It has one outstanding feature: the horizontal stripes on its legs. It has also long, narrow hooves – the narrowest of any equid. Their unique structure allows the animals to be swift and surefooted in their rough, rocky habitat. The ears are large and bordered by black colour whilst the thick, upright mane is also black at the tip.

Grass is the favoured food of the Somali Wild Ass, but they also feeds on shrubs and other desert plants. The animal grazes mostly when it is cooler – at dawn, dusk, and during the night. During the heat of the day, they often retreat to rocky hills to rest in shady spots. Given the hot environment it inhabits, there is no surprise that Somali Wild Ass stays within easy reach of water. They generally do not wander more than 32km from a drinking source. They can go without water longer than other equids, but they still need to drink at least once every two or three days.

Mares tend to produce 1 foal every 2 years and births occur during the wet season.

The Somali wild ass is critically endangered. This means they face an extremely high risk of extinction in the wild. A protected population of the Somali wild exists in zoos in Europe, North America and the Middle East.

Wild populations have declined for a number of reasons. For one thing, local people hunt the ass for food and for use in traditional medicine. Some people believe the animal's fat is an effective treatment against tuberculosis[6]. Hunting and drought has taken a greater toll in recent years, as political unrest in the region has allowed better access to automatic weapons. Other problems include increase of human populations and the expansion of settlements. Moreover, wild assess are competing with domestic livestock for limited grazing grounds and water sources. And as the wild and domestic animals (specifically donkeys) come into contact, there is more and more interbreeding – another serious threat to the wild ass.

It is widely believed that the Somali Wild Ass is on the brink of extinction in its wild habitat. The last few herds in

---

[6] In one of my trips to the mountainous areas of Sanaag, (November 2009), probably the last refuge of any remaining animal, I was informed by a health worker based in Gudmo Biyo-Cas that the prevalence of tuberculosis in the area is one of the highest in the country. (Author)

Somaliland are believed to have been either killed, or died of thirst during the 2004-2005 droughts in Sanaag and Sool regions. However, as late as 2006, the presence of a small population not more than half a dozen animals were reported in an area near Sha'ab village, 50 km north of Huluul town in western Sanaag.[7]

The Somali Wild Ass needs help if it is going to survive in the wild.

An essential first step is surveying the wild populations, if they remain in the wild, in order to know their numbers and distribution. This should be coupled with awareness raising on the protection of the species and extending humanitarian and developmental interventions to the communities in these areas, so as to discourage them from hunting the remaining animals.

---

[7] It was reported to me by Ali Abdigir, a Somali writer and researcher, that in 1999, one of the agenda points of a session in Puntland Parliament focused on the critical situation of the Wild Ass. According to Ali, the report was presented by Hussein Ismail, the first minister of the Range, Forests, Livestock and Agriculture, and unveiled that there was only 13 individual animals left in the wild. The report also added that the critically engendered animal devised night grazing as a new survival mechanism.

6

The Vanishing Trees of Hargeysa

O Allah! May we not sojourn in a barren, treeless place! (The Author)

Importance of Trees

The importance of trees becomes more evident upon imagination of a picture of a world without them. Cleaner air, water, food for our tables, and thoughts, as well as an inspiration for our senses are, but few of the multitudes of things we obtain from trees. If nothing else, they provide at least a place to sit and relax, close our eyes, and listen to the winds rustle through their leaves.

Here is a summary of the benefits of trees:

Trees slow down runoff by holding or absorbing water and improve moisture regime; trees provide shelter for people, animals and other plants; trees reduce the devastating effects of flooding. Trees provide nutrients and shelter for variety of organisms upon decomposition. Trees protect the world's climate by absorbing carbon dioxide *(carbon sinks)* and transpiring oxygen are often called 'lungs of the earth'. Trees provide certain habitats with the stable conditions necessary for life, and offer beautiful scenery. Trees prevent erosion and reduce water evaporation. Trees provide us with nourishment. Trees are constant source of medicine. Trees provide us with fuel wood, charcoal and coal.

Background

A little over a century ago, the valley of Hargeysa, also called *'Togga Maroodi Jeex'*, had been a heavily forested area in Somaliland with diverse riches of fauna and flora.[8] One of the earliest recorded descriptions of Hargeysa were captured in the pages of some of the books written by British travellers and big game hunters such as C. Peel and H.G.C. Swayne who made their expeditions to Somaliland in the late 1890's. The area used to have a wide variety of wildlife and plant species. It constituted a home for carnivores such as lions, leopards. The presence of these wild animals, attracted by the availability of water at the *Togga* and the thick forest cover, had kept pastoralists at bay during most parts of the year – only converging at the water points dotted along the seasonal watercourse during the *Jiilaal* season. The dominant tree species in the valley of Hargeysa were *Galool (A. bussei)*, *Qudhac (A. spirocarba)*, *Bilcil (A. mellifera)*, *Sogsog (A. etbaica)*, *Gob (Zizyphus mauritiana)*, *Cadaad (A. Senegal)* and many others. Major H. Rayne, a British administrator who served as the District Commissioner of Zeila wrote the following after visiting Hargeysa in the year 1921:

> "... Hargeysa, the usual African upcountry town, there is the District Commissioner (D.C.) Court, the police lines, the prison and the D.C.'s house. Beyond them (lied) the native town; a town of sticks, straw and native mats, with a few sun-dried brick houses, and one of stone building. Between all this is a natural park; a park on the banks of a waterless river-bed; a park filled with the thorn trees, the aloes in flower, and the plant with the thousands of

---

[8] Another name of Hargeysa seasonal watercourse, but not heard, found in the pages of some of the books written by early European travellers, is Aleya deera? See Major Swayne's *"Seventeen Trips through Somaliland"*. P 100.

fingers.[9] Strangely enough, there is a European women as well, who sits beneath a huge mimosa tree and put them all into a picture. She says they are beautiful".[10]

Until 1891, there was a small settlement at *Jameeca Weyn* part of Hargeysa, on the southern bank of the seasonal watercourse consisting of a number of mud houses and Somali huts *(Aqals)*. The construction of the residence of Lord Delemare was the first house built for a European in Hargeysa. Delemare made his first trip to Africa in 1891 to hunt lions in Somaliland, and returned yearly to resume the hunt. In 1894, he was severely mauled by an attacking lion, and was saved when his Somali gun bearer, Abdullahi Ashur, leaped on the lion, giving Delemare time to retrieve his rifle. As a result of the attack, Lord Delemare limped for the rest of his life; he also developed a healthy respect for Somalis.[11]

During the British administration in Somaliland, forest protection practices were adopted and water run-off was put under control by rock check dams and soil bunds. Cutting of trees was punishable by fine and/or arrest, while pound fee rates were imposed on the owners of animals found grazing in the town.[12]

---

[9] The tree that Major Rayne described as having "thousands of thorns" is most likely to be *Euphorbia nubica {"Ciin"}* which is still abundant in Hargeysa, particularly on either side of the dry river bed.
[10] M.C. Major H. Rayne, *Sun, Sand and Somalis* (1921)
[11] Bull, Bartle. (1992). *Safari: A Chronicle of Adventure*, p. 188.
[12] Pound: A public enclosure for stray or unlicensed animals existed in the site of Hadh-wanaag Hotel.

The Plight of trees in Hargeysa

The cause of the dramatic decrease in the number of trees in the city over the years can be attributed to the following reasons:

1. Urban densification is one significant cause of continuing tree loss in an urban centre. Over the years of its existence, Hargeysa has expanded from a small dwelling place at *Jamaaca Weyn* (1891) to a city sprawling out over 250 square kilometres. This expanse of land, particularly the valley, which used to be thickly wooded area, is being permanently converted into buildings, roads and pavements. The deforestation of Hargeysa valley started early, over a century ago, when the first permanent settlement of the city was made possible by Sheikh Madar, whereby his followers (*xertiisi*) began clearing the northern bank of the Togga for sorghum cultivation.
2. Soil compaction as a result of human activities, particularly movement of equipment, such as motor vehicles has reduced the permeability capacity of the soil, thus depriving the vegetation of water and other essential nutrients. This has also led to the heightening of water runoff along and over the sloping contour of the city, causing continuous infrastructural damage on roads and flood drainages. The thinning of trees in the valley caused the interruption in the nutrient cycle processes and the subsequent reduction in water infiltration into the ground.
3. There has been an indiscriminate cutting of trees for fencing, firewood and poles by the residents of Hargeysa. So far, the *Sha'ab* zone, which used to be mostly a public land, is the worst affected area, as it hosts a substantial number of internally displaced persons (IDPs) and returnees. It follows that common property is often subject to neglect and over-utilization.

4. The *Acacia bussei (galool)*, one of the most useful trees in the country and the dominant species of Hargeysa valley, has been a subject to continuous cutting of the outer covering of the woody stems (de-barking) for its *Asal*, often used in traditional medicine for its curative powers.[13] The de-barking processes deprive trees of the steady flow and transport of water and nutrients - ingredients vital to the healthy growth of trees.
5. Although the regeneration of the indigenous tree species is very slow and limited, the presence of great numbers of goats and camels in the city has its adverse effect on the growth of new plants. This has also a discouraging effect on any reforestation programme to be executed in the city.
6. There is a low environmental consciousness among government planners and the private sector. The sight of an old Acacia tree being foolishly bulldozed for the purpose of road clearance and/or construction project is becoming a normal practice. The idea of creating parks is alien to the Municipal officers, let alone protecting the remnant trees in the city from destruction, the wanton greed of land developers and the axe of the economically marginalized returnees who chop trees for fire-wood and fencing.
7. Absence of tree planting programmes in the city, particularly in the business centre where it became habitual for some members of the business community to cut trees in front of their buildings on the pretext of discouraging squatters establishing petty businesses such as teashops and kiosks under the trees (see plate # 6). There is no government mechanism to protect

---

[13] The *Asal* of the Galool is believed to be effective against diarrhoea. Use of *Asal* is heightened during outbreaks of cholera and acute watery diarrhoea.

these trees from abuse and cutting, and bring the culprits before justice.[14]
8. The 'choking' effect of plastic bags on Acacia trees known in the local parlance called *"the flowers of Hargeysa"* is believed to slow down the vital process of photosynthesis and make the trees under constant stress! The plastic bags are also an eyesore.
9. The habit of changing and disposal of engine oils in the open streets, mainly by bus drivers and truckers.

Future outlook and conservation efforts

Generally, the disappearance of trees bears dire environmental consequences in any ecosystem. Elderly residents of Hargeysa can be nostalgic about the waning of tree resources. No more 'whistling' of *Galool* thorns *(foorida galoolka)*[15] during the windy summer days and their yellowish sweet smelling flowers *(manka)* in the late autumn and summer! The worst scenario is a city devoid of vegetation but crowded with a different 'forests made up of buildings, machinery and garbage!

The combination of above factors have caused the temperature of the city to rise and the air to turn stale, in the absence of the cooling effects of plants and the fact that the very air we breathe is improved by the presence of trees. In order to feed themselves, trees absorb harmful

---

[14] For further elucidation on this matter, one can read the first essay of this book titled: "What happened to the tree that "thought" it was saved from barrenness (childlessness)!"

[15] This whistle is created by dried *Galool* fruits with holes. Summer winds make them whistle. The holes are made by fruit insects that lay their eggs inside the fruit when it is in the ripening stage. The fruit forms a safe habitat and food is available inside it as well. Once the insect reaches maturity, it perforates a hole and goes out of the dry fruit. Dried Acacia fruits of this kind are like deserted homes; however, they become a shelter for other insects.

chemicals such as carbon dioxide and in turn give off oxygen.[16]

Moreover, they filter and trap pollutants such as smoke, dust, and ash making our air cleaner.

The concept of conservation, protection of trees and propagating them is alien to the concerned local authorities. *Bilicda Magaalda* (city beautification) has become a buzzword for every Mayor of Hargeysa, but it is difficult to grasp how a city without trees can be regarded beautiful.

It is worth noting that the theme of 2005 World Environmental Day (WED) was 'Green Cities'. The United Nations conference held in San Francisco during the first week of June 2005 attracted a big number of mayors of world cities as the conference was dedicated to adopting sound environmental practices for urban centres – where the majority of world's population live.

It is never too late! It is a high time that the Mayors of the cities and towns should adopt sound environmental practices. And as a first move, the following recommendations need to be considered:

a) That the on-going, indiscriminate cutting of trees in the cities and towns be curbed.
b) Impose a directive that prior to building new houses, the owner(s) must first sign a pledge to plant at least two trees per plot;

---

[16] Carbon dioxide in normal quantities is not harmful, but essential for life on Earth. Only in increasing quantities it contributes to global warming

c) That a portion of the land under the jurisdiction of the municipalities is declared as protected area and to set aside about one acre of land within the township, as a park - where future tree planting activities can be carried out.
d) Management and proper disposal of the garbage that has almost engulfed all urban centres
e) Public awareness creation on the protection of urban environments

7

Denudation of Batalaale Conocarpus lancifolius Plantation

Batalaale is piece of coastal land just to the north of Berbera town proper. Its eastern fringe reaches the elevated ground of coral stones, where the uncompleted Russian-built hospital stands. To the north, it is bordered by the sea and old Bender Abbaas, the original site of Berbera, with its huge cemetery[17] at the sea-strand.

Batalaale is very popular among the Somalis as the final resting place of the legendary Somali poet, Elmi Boodheri, the first and the only person known to have died of love. One line from Ahmed Ali Egal's popular song *"Boodheri"* immortalizing his fate runs like this:

> "Don't you know that the sea is dark[18] (meaning blue),
> And that it is composed of water!?
> Don't you know that Love laid down Boodheri at Batalaale beach (as his final resting place)?
> Don't you know that well-wishers pay him respect at his grave?"

In nutshell, the meaning of the above few lines is: *"As it is an undisputed fact that the sea is dark and that it is a body of water; equally it is a fact that one could die for the sake of love"*.

---

[17] "Xabaalaha Saadada" (The Graves of the Elect) is referred to local resting place of Saints from Arabia.

[18] The color of the sea is blue, however, and strange enough, there is no Somali term for blue colour.

Batalaale plantation was established in late 1945, to serve as a windbreak and to stimulate the weather of Berbera during the hot summer days and for its aesthetic value. This was a good start initiated by J. J. Lorry, a British forester from the Department of Natural Resources (DNR). Prior to the afforestation programme, the area was dominated by *Suaeda fruticosa* and other salsolaceous salt tolerant bushes, shrubs, and grasses such as *Darif (Lasiurus hirsutus)* and *Dungaari (Panicum turgidum)*. A wide scale planting of Doum palm *(Hyphaene thebaica)*, date palm and *Dhamas (Conocarpus lancifolius)* was carried out during the period 1945-1950 with an impressive output of 12,000 trees (raised in the local nursery of Berbera).[19] This was following by another large scale planting of *Conocarpus lancifolius* from 1975-78. Moreover numerous shallow wells were dug within the site for watering the trees.

After the collapse of Siad Barre's regime and return of people to the area, many returnees and internally displaced persons (IDPs) who settled Berbera town began cutting the trees for shelter material and firewood. The habit of keeping camels and shoats by some residents and indiscriminately lopping tree branches in order to feed animals had its deleterious effect on the micro-forest of Batalaale. It thus makes any effort to reverse the trend almost impossible. The stumps of the cut trees can now be seen in rows – a reminder of the cruel deforestation that took place in the area. Moreover, even the location of the grave of late Boodheri is nowhere to be found.

There is now a small reforestation trial site surrounded with wire fencing to ward off the browsers. Interestingly,

---

[19] *Dhamas (Conocarpus lancifolius)*, an indigenous tree endemic to northern coastal areas of the Somali territory. It is habitat is along the banks of northern coastal seasonal water courses

the wider area is now being colonized by Mesquite (*Prosopis juliflora*)[20]. Unfortunately, the cleared land attracted the attention of greedy land speculators and equally some irresponsible government officials. Consequently, parceling of the land into plots for private use has begun in a big scale.

---

[20] See the next chapter on Mesquite.

Mesquite (Prosopis juliflora)[21]

> "If you cannot win the war on mesquite, better reconcile with it, and make the maximum benefit out of it. This can be an effective way to check its unnecessary spreading".
> (The author)

Prosopis juliflora is a perennial deciduous thorny shrub or small tree that can grow up to the height of 6-7m. It is an evergreen plant native to northern South America, Central America and the Caribbean. It is fast growing, nitrogen-fixing and tolerant to arid conditions and saline soils.

This tree is recognized for its aggressive growth, and wide distribution in many parts of the world. It is believed to be an invader and noxious plant that encroaches on all available land. It has a massive root system that can grow down as far as 15 m deep to draw water and occasionally more than double that depth. Its canopy and the large quantities of leaf litter create a condition that does not allow other to germinate below the shade.

A near relative, *Prosopis cineraria* (native of India) have been shown to reach a depth of 60m in the Oman desert (Le Houerou) in order to reach the water table.

In Somaliland, the plant is known as '*Garanwaa*' literally meaning 'the unknown' – so for its abrupt and massive recruitment into the country that probably did not give the local people a chance to give it a fitting name. In the South

---

[21] The article was abridged from a research document:
"*Proliferation of Honey Mesquite in Somaliland: Opportunities and Challenges*", (March 2006), Ahmed Ibrahim Awale and Ahmed Jama Sugulle, Candlelight Org.

central zones, it is *"Geed Yuhuud"* (the tree of the Jews[22]), *"Geed Jinni"*, the Tree of the Devil and *Ali Garoob*.

Native Americans still use mesquite pods as a staple food. They make tea, syrup, and ground meal called *Pinole* from various parts of the tree. Mesquite bark is used for making baskets, fabrics and medicine. It is so useful to them that it can be compared, in terms of utilization, to our *Galool (Acacia bussei)*. The wood of the mesquite tree burns slowly but with great heat[23] and is used as firewood, for its aromatic luscious flavor it imparts to grilled meats and fish.

In many parts of the world, products made from mesquite are on the rise. Mesquite has a wonderful wood that is ideal for high quality furniture – for it undergoes very little shrinkage or swelling.

Mesquite was first introduced in Somaliland as early as 1950 in Bullaxaar (Bulhar) town, on the Gulf of Aden coast by a British forester, V.H.W. Dowson, for use as a windbreak for a date palm plantation project – also introduced to that location during the same year. Three decades later, the plant became a prime choice for some development organizations involved in the support of the refugees displaced by the 1977 Ethio-Somali war – better known as the Ogaden War – for a reforestation programme in and around the refugee camps scattered then in

---

[22] Reference is made here to the famous Gharqad tree mentioned in In Sahih Muslim: Book 041, Number 6985

[23] 61 Mesquite fire reaches temperatures as high as 8,050 Btu/lb.

Northern Somalia. The introduction of mesquite also continued at that time in the South.

Mesquite stands are now well-established along the banks of seasonal water courses (*Togga*) replacing local species and forming dense thickets – simply because of its prolific growth and under-utilization compared to over-utilization of local species. Mesquite is also the preferred urban tree species for shade. Its 'virtues' were summed up by an elderly women residing in one of the returnee re-settlement areas of Hargeysa who, one day, approached Candlelight's natural resource management (NRM) team, during one of their tree distribution trips, in order to receive a seedling. To the surprise of the team members, she requested a *Garanwaa* (*P. juliflora*) seedling, whereby after the surprised team members wanted to hear more about her liking of the plant. They wondered 'why should almost everyone else treat mesquite as an accursed plant, whereas this woman regards it as her first choice?' Her response was simple and clear: "*Mesquite gives good shade; one doesn't need to worry about watering it or protecting it from goats or even humans for toothbrush sticks (caday), and it even does not need fencing. It is fast growing and long lasting as well!*"

Few years later, the tree grew big, with broad and spreading horizontal branches, providing good shade to the women and children in her neighbourhood. Not only her efforts were appreciated, but also her status within her community grew.

Notwithstanding the above mentioned virtues, there has been a growing criticism among local communities calling for its eradication. This means that its multi-faceted benefits are often eclipsed by its negative attributes and characteristics. Some of these characteristics include its dense and impenetrable formation with its unfavourable socio-economic impacts, its competition with the indigenous species often resulting in the suppression of its competitors, and henceforth, loss in biodiversity. The grazing potential of the rangelands is also reduced through the process. Due to the high sugar content in its pods, tooth decay affecting livestock accompanied by stomach disorders results in the reduction of feed intake, and hence, their emaciation and succumbing to diseases. Its deep root system draws more than its fair share of available moisture, making it difficult for other plants to survive beneath its canopy or around it.

On the other hand, there is a growing consumption of the plant in urban areas for charcoal, firewood, shelter material, fencing (even though its branches are not as durable as many indigenous trees), etc. Therefore, this issue surrounding the plant merits a much closer study, at a time the natural vegetation of the country is becoming progressively depleted, whereas the demand for biomass energy is increasing day after day.

A sure management strategy to control the speedy spreading of the plant is by maximizing its utilization for different purposes. Uprooting young plants in their early growth stage is an effective method used for controlling its prolific spreading into farmlands. Firing as well as stumping plants at 10cm below the ground in order to eliminate the chance of re-sprouting is also an effective technique. The use of used engine oils for its eradication has its negative environmental consequences on soil and other living organisms around the plant. However, local experience and elsewhere shows that the war against mesquite is not only expensive but also almost unwinnable. Therefore, one can reasonably say; *"If you cannot win the war on mesquite, better reconcile with it, and make the maximum benefit out of it. This can be an effective way to check its unnecessary spreading"*.

9

Deforestation of the Mountains of Sheikh

> "Dulmi qabe hidhiilow, muxuu dogobbo moofeeye,
> Isagoo naftiisii dulmiyey, deynna lagu yeehsay,
> Dantiisaba ma raacee muxuu soo dab-dacaleeyay
> Goor waagu daalacay muxuu leydhka dib u eegay"
> __ *Tixdan lama yaqaan cid tirisay (anonymous).*

Oh Hidhiile[24], a culprit you are!
Many are the tree trunks that were put into kiln by you.
While your action was an utter transgression against your soul,
You are also in a situation where you are always drowning in debt.
Living in isolation and removed from family and friends and unable to detach yourself from such sinister business.
In the morning hours you are always on the lookout for the headlight of the charcoal businessman's truck to arrive and replenish your needs – mainly Qat and rations.[25]
____ *(anonymous poem translated by the author from the Somali)*

Charcoal production activities in the *Ogo* areas has been going on now for decades and, therefore, has resulted in the near depletion of the traditionally *Acacia bussei* rich

---

[24] *Hidhiile:* a small scale charcoal producer, who often operates singularly.
[25] *Qat: Catha edulis*: and evergreen shrub, whose leaves are chewed mostly by men that cause euphoric effects, is becoming popular throughout the area as a cash crop due to its high demand by both urban and rural people.

woodlands towards the south of Golis Range. This caused charcoal producers to expand their areas of operations towards the watershed areas of the Golis Range.

Deforestation of watershed areas will definitely result in serious consequences that could negatively affect the quality of life of the communities who derive their livelihoods from the mountain areas as well as those living in the lowlands.

One of the most serious problems that could result from the denudation of watershed areas is the decrease of water infiltration capacity of mountains causing the few remaining streams to dry up.

This could also mean dramatic fall in water recharge of shallow wells, increase of water salinity, over-silting of dams and general decrease in agricultural production. Similarly, there could also be an increase of water runoff causing watercourse bank erosion as well as human and material losses. These problems will also trigger human displacement and their movement to new areas.

One particular area of interest in the Golis Range is the mountain side of the arterial Sheikh Pass of Sahil Region (Photo # 17). Nowadays, many travellers passing this area can easily see an escalation of charcoal production activities being carried out by pastoralists whereby bags of charcoal ready for sale are lined along the sides of the tarmac road. The sharp rise in the price of food on one hand and the increasing demand for charcoal in the urban centres is the driving force behind this trend and encouraging pastoralists to supplement charcoal production to livestock rearing.

Some of the effects of deforestation on watershed include:

- **Destruction of forests** leads to tragic loss of biodiversity.
- **Heavy soil erosion** caused by accelerated water runoff.
- **Flooding** – one major importance of trees is that they allow water to infiltrate into the ground during heavy rains. Loss of tress means less water is absorbed into the ground.
- **Landslides** - The roots of the trees bind soil to it and to the bedrock underlying it. That is how trees prevent soil from getting eroded by natural agents like wind or water. When trees are uprooted, there will be nothing to hold the soil together thus increasing the risk for landslides that can cause seriously threaten the safety of the people and damage their properties. Trees also slow down the speed of descending downdrafts and its effects on the lowlands.

It has been observed that there is an increase of falling rocks and sand particles onto the tarmac road that sometimes blocks the road, or causes traffic accidents.

The February 2006 landslide in the Philippines triggered by heavy rains buried a complete village whereby its 300 houses were covered by mudslide. The cause of the tragedy was the deforestation of the watershed areas overlooking the village.

- **Degraded Watershed** – This leads to loss of sustainable water supplies for lowland communities. This is because trees affect the hydrological cycle. They can change the amount of water in the soil, groundwater, and in the atmosphere.

- Deforestation will also degrade the tourist potentiality of Sheikh Area.

If the current trend with its destructive effects is not addressed by the concerned authorities (particularly the district authority of Sheikh town and the Ministry of Public Works) it will be unavoidable to experience these above mentioned dire consequences in the near future.

10

Charcoal Production—Great Threat to Somali Livelihoods

"Whenever I see trees on fire, I sense a smell of heartbreak in the air". ___ the author.

The Somali region, like other semi-arid areas in the world, has been in a state of ecological change, for many decades and perhaps for hundreds of years, and practically all the changes have been towards environmental degradation. However, the pace of this process during the past few decades has reached levels not paralleled in the recorded history (in paper and in memory) of the Somali region. Some of the most visible impacts are the dramatic decrease in vegetation cover and decline in the prolificacy of wild life. Charcoal production, wood for construction and establishment of thorn enclosures are some of the main factor contributing to deforestation.

Biomass is the main traditional source of energy for the Somali population. Charcoal is the principal energy producing fuel commonly used in urban areas for cooking and heating whereas firewood is common in the rural settlements.

The best charcoal is prepared from acacia species that have great socio-economic and environmental benefits. The selective cutting of the *galool* (*Acacia bussei*), the most preferred tree for charcoal production, has led to its over-exploitation. The rate of deforestation has already outpaced the regeneration capacity of the remaining woodlands.

Moreover, indigenous species are slow growing and it takes a considerable number of years (average 25-30 years) to be mature enough (depending on the amount of moisture received) to produce four sacks of charcoal (equivalent 80 kg).

Following the disintegration of the Somali state in 1991, the exportation of charcoal to the Gulf States, particularly from Puntland and South/Central Somalia, has turned into a highly lucrative business. The main ports reputed for charcoal exportation were Ceelaayo, Boosaaso, Ceelmacaan, Marka, Baraawe and Kismaayo. Concerted efforts made by Puntland authorities and the civil society organizations (CSOs) resulted in the ban of charcoal export, and considering it as an illicit trade. Hike in charcoal price within Puntland markets, decline in the number of trees and the growing local demand were some of the factors that contributed to the ban of charcoal export from ports within Puntland jurisdiction. However, in the south-central zone, which remained unstable for many years, charcoal production and exportation became an important factor in the war economy that was used by the powerful to fuel the civil war. Thus, the Southern Somalia ports remained important charcoal exit routes. This is a good indicator of the connection between chaotic situations, weak governance or its absence, on one hand, and the extent of environmental damage resulting from these situations, on the other.

The extent of the vegetation loss can be clearly understood from statistics collected in 2003 by a civil society organization (CSO) based in Kismaayo. During that year, a total of 85,970.50 tons of charcoal (Photo # 16) destined for the Gulf States were shipped from Kismaayo port on 44 ships and 59 dhows. *{Source: KISIMA Peace and Development Organization, Kismaayo}.*

By denuding the long-growing hard wood species of the country's rangelands, the massive charcoal trade leads to a number of severe environmental consequences including:

- Loss of forage for livestock;

- Loss of water retention and moisture regime in soils resulting from soil compaction (hard pan);

- Unfavourable changes in soil conditions, micro-climates, disruption in rainfall, etc.

- Reduction in the amounts of carbon sequestered in plants as the capacity of trees to serve as carbon sinks is diminished. This will result more greenhouse gases in the atmosphere leading to global warming and its effects;

- Increased sheet and gulley erosion;

- Loss of woodlands and a number of important forest products, and;

- Loss of biodiversity.

Given that Somalia's domestic economic activity is dominated by livestock production, and the Somali cultural core is oriented around the pastoral way of life, this unabated destruction of the vegetation cover is tantamount to a direct and long term assault on the quality of life of Somali people. It is high time that measures leading to the reduction on dependence on charcoal on one hand, and popularization the utilization of other cooking energy sources such as kerosene, liquefied petroleum gas (LPG) and solar cookers on the other be adopted in order to reduce the scale of deforestation.

11

Climate Change

> "Corruption has appeared throughout the land and sea by (reason of) what the hands of people have earned, so He (i.e. Allah) may let them taste part of (the consequences of) what they have done that perhaps they will return (to righteousness)." __Holy Qur'aan:30:41

Climate change is a study dealing with the variations in climate on many different time scales and the possible causes of such variations, whether due to natural variability or as a result of anthropogenic factors (human activities). Equally, global warming is the gradual increase of temperature in the lower atmosphere of the earth due to the accumulation of greenhouse gases. The term "greenhouse" was coined to denote buildings with glass or plastic walls and a roof supported by metal framework used for the cultivation of vegetables and flowers – particularly in cold weather conditions.

Here is how such a greenhouse works: When sunlight strikes the glass, most of it passes through and warms up the air, soil and plants in the greenhouse. These objects give off heat, but the returning heat waves, in the form of infra-red radiation, have a longer wavelength than the incoming rays of the sun[26]. The long wave (returning) radiation cannot pass through the glass as easily as the

---

[26] Energy intensity is inversely proportional to its wavelength: Long wavelengths have low energy, while short wave lengths have high energy.

short wave, and is re-radiated in the greenhouse causing everything to heat up.

In a similar manner, sunlight, which is a short wave radiation, passes quite easily through the earth atmosphere. When it strikes the earth, heat is given off and the surface is heated. Long wave radiation is given off which is radiated back into the atmosphere. Some of this long wave radiation escapes into space, but most of it is absorbed by carbon dioxide, plus other gases that exist in small quantities in the atmosphere. These gases form a "blanket" that keeps the temperature at an average of 15ºC. Without the natural greenhouse effect, Earth' surface temperature would be -18ºC. The accumulations of greenhouse gases (including carbon dioxide) are responsible for the warming up of earth's atmosphere and altering its energy balance. However, carbon dioxide is vital to life on earth. Plants absorb it to convert sunlight into food. This is called photosynthesis. In this process plants give off oxygen. In the absence of carbon dioxide, plants will not be green, oxygen will not be produced, and eventually, life on earth would not have been possible. The gases in the earth's atmosphere consist of 78% nitrogen, 21% oxygen, while the remaining 1% consists of several gases - the most important being carbon dioxide amounting to 0.03%.

The alteration in the chemistry of the atmosphere and changes in the global temperature, although apparently small, could cause very large change in climate.

In short, climate change is caused by:

a) Concentration of greenhouse gases in the atmosphere – and especially carbon dioxide – that have risen over the past two hundred and fifty years, largely due to the combustion of fossil fuels for energy production. Concentrations of Methane have also risen due to cattle production, the cultivation of rice, and release from landfills;

b) Land-use change: When ecosystems are altered and vegetation is either burned or removed, the carbon stored in them is released to the atmosphere as carbon dioxide. Land use changes also alter surface albedo and the rate of absorption of heat.

The Inter-governmental Panel on Climate Change (IPCC) foresees temperature rise of 1.4 to 5.8 degrees Celsius during this century. This will definitely have dire consequences on the life on earth. Some of the effects will be:

- The melting of the arctic permafrost and Antarctic ice could dramatically worsen global warming by releasing massive amounts of greenhouse gases (most importantly methane) into the atmosphere;
- Rising sea levels and inundation of lowlands: Some of the countries that are at high risk of being inundated are The Netherlands, Bangladesh, and many Island nations in the Pacific and the Indian Ocean.
- Extreme weather conditions (increase in the intensity and frequency of flooding, droughts, high temperatures, wild fires etc.)
- Coral bleaching resulting from the photosynthetic symbionts of corals (*zooxanthellae*) becoming increasingly vulnerable to damage by light at higher than normal temperatures. The resulting damage leads to the expulsion of important organisms (algae) from

the coral host. Corals then tend to die in great numbers, thus negatively impacting the bio-diversity of marine life.
- Change in the distribution of disease-bearing organisms and increase in the transmission of many infectious diseases, including malaria, dengue, and yellow fever extending to temperate zones;
- Regional change in crop yields and productivity;
- Increased risk of famines, particularly in subtropical and tropical in arid and semi-arid lands (ASAL);
- Shifts in river flow and water supply, with serious implications for human settlements and agriculture;
- The depletion of the ozone layer. The ozone layer helps to shield life on earth from the damaging ultraviolet radiation of the sun;
- Rapid extinction of plant and animal species.

In the same token, some of the visible effects of climate change that can be experienced in the region are:

- Remarkable increase of temperatures in higher altitude ecological zones compared to earlier years, specifically in the major towns such Hargeysa, Burco and Boorama. The seemingly disappearance of some Somali terms used in expressing a very cold weather is another indicator of climate processes are in the making. Words like "*gabadano, gawre* and *juube*" were used to denote severe cold conditions. There is also a marked decline in the popularity and sale of second hand cold weather clothes – locally known as "*Huu-dhayd*", as there is less need for them as before.
- Decrease in precipitation levels, particularly in the mist forest areas of the country. The decrease in mist levels

is the main cause of the decadence and mortality of the Juniper forests (*Juniperus procera*) in Gacan Libaax and some other higher altitude areas in the country;

- Changes in the biological succession of some of the plants in certain vegetation zones. For example, the unexplained high mortality of *Acacia tortilis* {Qudhac} and *Balanites orbicularis* {Kulan} in the *Guban* is as is a disturbing trend[27];
- Extreme weather conditions (severe droughts, flash floods and heavy windy rains).

The biggest contributors of emissions that lead to global warming are the rich and developed nations of the world while the poor nations are those who bear the brunt of the dire consequence of climate change. However, this does not relieve everyone from the common and shared responsibility of caring for the planet and to wake up to the fact that, if our planet suffers, all of us will suffer too and we have nowhere else to go. It is a high time that we came to realize the negative impact of our actions and stop corrupting our planet, if we have to avoid the terrible consequences.

In dealing with climate change, two different, but inter-related tasks has to be adopted:

Adaptation: This means adjustment in natural or human systems in response to actual or expected climatic stimuli or their effects, which moderates harm or exploits beneficial opportunities.

Mitigation: Climate change mitigation may be defined as any attempt to reduce the rate at which greenhouse gases are accumulating in the atmosphere. The two options that

---

[27] For further elucidation, the article titled "the Mysterious High Mortality of Acacia Trees in the Coastal Areas" can be read.

can be considered are: Reducing emissions and carbon sequestration i.e. processes that remove carbon dioxide from the atmosphere.

Finally, some of the means to confront the threat of climate change are:

➢ Framework of international/ national environmental law;
➢ Environmental compliance[28], enforcement and liability[29];
➢ Adoption of new resource-efficient (clean) technologies;
➢ Environmental consideration in decision-making;
➢ Preventive action, concerted response, governance;
➢ Awareness raising, education, capacity development;
➢ Cooperation of all actors for a sustainable future.

---

[28] **Environmental Compliance** means conforming to environmental laws, regulations, standards and other requirements.

[29] **Environmental liability:** is a term used for the process whereby responsibility for the cost of damaging the environment is transferred back to those that cause the damage. The principle under which **environmental liability** operates is sometimes called 'the polluter pays principle' and the ultimate objective is to reduce damage to the environment.

12

Al Wabra Wildlife Preservation (AWWP)

> For children to kill a lizard is a pastime, while it is a loss of life for the lizard __ Somali Proverb
>
> If a man kills a lion they call it an act of bravery, but if a lion kills a man they call it a beastly act. __ (Anonymous)

The Al Wabra farm is an oasis of green areas, palm trees and many rare wild animals from all over the world. Situated in central Qatar, its website informs the public that it is invested by Sheikh Saoud Bin Mohammed Bin Ali Al-Thani's passion for nature, and is a base for an international team of expert vets, biologists and keepers who are dedicated to the care and conservation of the rare and mostly endangered animals from different parts of the world. Among thousands of animal members in AWWP, there are a number of critically endangered species from the Horn of Africa, particularly the Somali region – now cared for in AWWP. These include the endemic Somali Wild Ass (*Equus africanus somaliensis*), Beira Antelope (*Dorcatragus megalotis*), Dibtag Antelope (*Ammoodorcas clarkei*), Swayne's hartebeest (*Alcelaphus buselaphus swaynei*) and many others.

It is noteworthy to mention that the Qatari prince, Sheikh Saoud and a team of wildlife experts were on a trip to Somaliland in the middle of 1995, with the consent of the local authorities, in order to catch some wildlife for AWWP. The mission had mobilized sophisticated equipment including helicopters for use in capturing the animals, and actually succeeded in collecting an unspecified number of wildlife species. However, the objective of the mission was misinterpreted and had sparked a heated environmental debate at national level and protest by local communities in Oodweyne and Caynabo districts who accused the group of being poachers. The operation was completely halted following a bloody conflict involving the communities in the target areas versus the Prince's team.

The Qatari prince denied the accusations, saying his objective was preservation of wildlife and not poaching. According to the British Broadcasting Corporation (BBC) (31.3.1995), during a meeting with the late president of Somaliland, Mohamed Ibrahim Egal, the prince promised, among other things, to sink a number of bore holes in the region for the benefit of the communities in exchange of the wildlife.

News published in the website of Al Wabra Wildlife Preservation reports that the animal species transported in 1995 have multiplying in number and are safe from human disturbance. Currently, these animals seem to be luckier than herds left behind that are under the threat of poaching and stress. However, the fact is that, they could have fared better in their original habitats, had there been an environment conducive for their existence, natural breeding and perpetuation. In any case, the possibility that some of these animals will return to their original habitats, when the environment is not hostile anymore for them to breed and multiply in peace, cannot be discounted.

13

High Mortality of Acacia Trees in the Coastal Areas

> "We did not inherit the earth from our ancestors, but borrowed it from our children" _____Mahatma Gandhi

Some years back, the phenomenal drying and dying of large numbers of Umbrella Thorn Acacia (*Acacia tortilis*) (see plate # 13) and to a lesser extent *Balanites orbicularis* (*'Kulan'*) in *Guban* coastal areas was the subject of discussion among the coastal communities.

One has only to look the roadside just outside Berbera, near the road checkpoint to see hundreds of young dead acacias whose dead stems and branches have disintegrated expansively onto the ground.

An article published in *Haatuf* newspaper (issue 1497, 6th August 2007) carried the following title: *"Dhimashada Geedweynta Xeebaha Duleedka Berbera waa Mushkilad u Baahan in Jawaab loo Helo"* (The death of coastal trees in the Vicinity of Berbera is a Problem which cries for explanation). The author of the article has exposed the deep concern expressed by some community members residing in Berbera as well as some pastoralists in the area. One pastoralist was quoted to have said:

> "No matter how severe a drought spell in the past used to be, we have not experienced anything like the current massive dying of these trees. A tree that was green a month ago is now dry and dead. You will see them shrivelling up within few days. This is a strange phenomenon and we are wondering what sort of calamity has befallen upon these trees. Even dry trees will not have

firewood as the dry branches crumble under the light pressure of your feet".

Community members interviewed attribute the cause of the premature drying of these trees to toxic waste which leaked from surface to air missiles (SAM-2 & SAM-3) left by the Soviets during late 1970's when the relationship between the then Somali Republic and the Soviet Union became strained as a result of war between Somalia and Ethiopia in 1977.

Although the issue under discussion calls for an in-depth investigation, the current author has a feeling that the cause could be different. The basis of this argument is:

a) The unusual mortality of the *Qudhac* and *Kulan* trees is not confined to the vicinity of Berbera. I witnessed a similar fate of these species in the coastal area south of Bullaxaar (Bulhar) village (approx. 60 km west of Berbera).

b) If the leaked chemicals were the causing the problem, one may ask why human beings and animals in the area were not affected.

c) Most of the trees that are affected are located in a belt that runs parallel with the sea (Gulf of Aden) with a breadth of approximately 5-10 km, but at the same time stretching hundreds of miles along the northern coastline.

No Chemicals but Climate Change is the Culprit

The death of the *Qudhac* and *Kulan* trees in that particular vegetation zone which runs approximately tens of kilometres along the coast can be attributed to the effects of climate change. Similar experiences have been noted in many parts of the world, particularly in the dry land areas.

The declining precipitation levels coupled with increase in temperatures over the past several decades in the higher altitude mist forest areas of the Golis Range, such as Gacan Libaax, has also resulted in the decline and mortality of the Juniper forests (*Juniperus procera*).

Whenever temperature levels rise, plants experience stressful conditions. The demand of trees for water proportionally increases as the rate of evatranspiration increases. Therefore, in the absence of sufficient supply of water to replace the amount released from the leaves, plants will end up in a stressful condition which could have a far-reaching effect on its chemistry and functioning, making them susceptible to plant diseases and pest infestations. Finally this leads to the eventual dying of the plant.

One can envisage the negative environmental and socio-economic impact of such large scale dying of those trees and the absence of recruitment of new trees to replace the dead ones.

Attaining a solution to this complex problem at a local level is not easy.

However, this can be the tip of an iceberg and a wakeup call alarming us an earth in crisis as well as similar dying and extinction of smaller and less noticeable living things around us. Therefore, a thorough investigation of the subject matter is a matter of urgency; while at least, understanding the root cause could be a part of the solution.

14

He who Plants a Tree, Plants Hope

> "If the Hour (doomsday) is imminent, and anyone of you has a palm shoot (to plant) in his/her hand, and is able to plant it before the Hour strikes, he/she should do so, because he/she will get a recompense (from Allah) for that action".____ A *Hadith*[30] by Prophet Muhammad

A quote from Lucy Larcom (poet, educator and editor in the 1800's) says: "He who plants a tree, plants hope". I pondered upon this quote and realized the great wisdom contained in it. Planting a tree links a person to both foreseeable and a distant future. Planting a tree is also one of the most rewarding activities a person could perform.

Planting a tree is a renewable source of reward. In one of his *Hadiths*, Prophet Muhammad (peace and blessing of Allah be upon him) said:

> There is none amongst the believers who plants a tree, or grows a seed, and then a bird, or a person, or an animal eats thereof, but it is regarded as having given a charitable gift {for which there is a great recompense}" {Al-Bukhari, III; 513}.

Planting a tree also sows patience in a person as it takes a long period of time from germination to maturity. When the intention of planting a tree crystallizes in someone's

---

[30] a tradition based on reports of the sayings and activities of Prophet Muhammad (PBUH)

mind, he/she is cognizant that it can be a long-term investment that can be compared to the process of procreation i.e. fathering/mothering a child in the hope that it may support and take care of the parents in their old age. Parents then will have to ensure that they nurture the child with the physical, emotional and spiritual needs.

Like a good son or a daughter, a tree never becomes ungrateful. This reminds me a story narrated to me by an old man from the riverine areas of southern Somalia whom I met, not long ago, in Hargeysa. He was displaced by the internecine civil war in 1990's, uprooting him from his home and farmland. He said while emphasizing on the importance of trees:

> 'a fifty-year old mango tree in my farm, which I planted with my own hands, remained reliably faithful to me till before I was torn from my farm. If you raise a child, it may desert you, while your mango tree remains close to you. I used to lie down just under it and wait for its fruits to fall upon me. All that I used to do was to extend my hand and relish eating it".

Verily! Planting a tree is sowing hope. In early 1991, having returned to Hargeysa from a refugee camp in eastern Ethiopia, just after a three-year civil war, I found Hargeysa like a ghost town. There was utter destruction wherever you turn your eyes to; while the sight of gutted, roofless and doorless buildings reminded me of an unclothed, raped and humiliated chaste girl. I used to roam and wander the empty streets and pathways, carrying with me a grieving heart. Then on one certain day and that certain moment, my eyes caught a breathtakingly beautiful boungavillea standing proudly amidst the rubble of a destroyed house with its crimson/red colour of its flowers – probably seemed in solidarity with the dead, the wounded and the shedding of human blood. The sight was comforting and refreshing that I had more energy throughout that day. The message it was putting across was as eloquent as any Somali orator. *"Truly with Hardship, comes Ease (thereafter".*[31] Hargeysa and its people shall prevail! From that day onwards, I developed a strong attachment to this plant.

One final note: I admire the community of Gebilay town for giving a new meaning and dimension to tree planting. It is said that every newlywed couple who perform their *"Nikah"* in Gebilay voluntarily plant a tree along the road between Gebilay and Arabsiyo, thus establishing a strong relationship between tree planting and human fertility. Some of the couples use the trees they had planted in the past as an indicator of the health of their relationship and as a love meter! If they notice that it is showing signs of stress or ill-health, an equal feeling resonates in their hearts and heads, sensing that is amiss with their relationship which requires urgent intervention. Then the ensuing concern urges them to provide extra care to the plant such as, watering, harrowing, and reinforcement of the thorn protection.

---

[31] *Holy Qur'an:* Surah Inshiraah 94

Declining Mist Levels over Gacan Libaax Mountain: Causes and Consequences

"The grass used to hold the mist settling on it; it nourished the shoots and then it seeped into the ground. The mist used to cover the area starting from *Asr* prayer time (in the afternoon) and continued to overshadow us till 9:00 a.m. on the following day. We used to call it **Hayays**, because of its incessant downpour. It dampened the soil and rejuvenated the vegetation at a time our livestock were in a dire need for pasture and browse".

____Xalwo Cige, the mother of the author who has spent most of her early years in Gacan Libaax

1. Introduction
1.1. *Description of physical and ecological context*

The site is Gacan Libaax Mountain (10^0N, 45_0E), circa 1718.90m above sea level. It is a part of an extensive highland ecosystem in central Somaliland running parallel with the Gulf of Aden.

The area is reputed for its gazetted juniper forest and other plant species, found also in some other areas sharing similar ecological characteristic. Hemming (1966) considered the Juniper forest of Gacan Libaax to be a climatic relict in a sense it may have survived from a period of greater rainfall in the past. Also in the work of Miskell (2000) which is the most recent ecological assessment in Gacan Libaax, he writes:

"It (the mountain) receives moisture in the form of mist, as maritime air is forced up over the area. This moisture helps to support patches of forest, which themselves facilitate the formation and entrapment of mist in self-contained system. If the forest is degraded the cycle may be broken..."

The other main evergreen trees and shrubs that are found with the

Juniperus procera ("Dayib") are *Buxus hilderbrandtii (Dhosoq), Euphorbia grandis (Xasaadin), Olea subtrinervata Chiov (Weger), Ficus sp. (Berde), Sideroxylon buxifolium (Shooy), Euclea schimperi (Maayeer), Cadia purpurea (Salamac), Acokanthera schimperi (Waabay), Dodonaea viscosa (Xayramad), and Draceana schizantha (Mooli)*[32] – the last is exclusively found hanging from the escarpment just below the rim on inaccessible areas. Some of the exotic species experimented and introduced by the British colonial forestry officers in the 1950s are *Cupressus sp.* And *Casuarina equisetifolia, Acacia Cyanophylla*, and *Eucalyptus camaldulensis (Baxra saaf)*. The last one, which is doing poorly, was introduced into the area in the 1970's.

1.2. Socio-economic context

In the past, pure pastoralism was the principal mode of production system in the area whereby the inhabitants along with their livestock (cattle, camels, sheep/goats) followed seasonal migration patterns, mainly north/south movement, depending on rainfall and pasture availability. However, since the last 30 years, the pattern of land use has drastically changed into sedentary agro-pastoral rain-dependent farming. The principal crops are sorghum and

---

[32] (Mooli (Draceana schizantha), *is also known in the eastern Sanaag and Puntland as "Dinaw". Ropes are prepared from its leaves, while wooden containers are crafted from its trunk.*

maize. *Qat (catha edulis)*[33], is also cultivated by some households as a cash crop due to its widespread addiction among the male population in the area, and elsewhere in the country as well.

The population in the catchment area of Gacan Libaax, that can be termed as the mountain resource users, including the surrounding villages can be estimated at 3,000 families (with an average family size of 7-8 persons), 53% women. The main villages in the area are Go'da Weyn, Go'da Yar, Deri Maraa, Iskudar and Biyofadhiisinka which serve as petty trading posts for the community. The valley between Go'da Yar and Go'da Weyn, the two main villages in the area is fertile and produces sorghum, maize, cowpeas and fodder.

The last one is transported and sold to livestock traders in the port town of Berbera for feeding animals during their shipment to the Arabian markets.

A variety of sisal (*Agave sisalana*), introduced into the area the late 1950's, is widely used as live fence. Each plant produces a single straight pole which is used as building material. Out of the fiber of its large leaves, women weave ropes and thus generate some income from doing so.

The decline in the livestock herd sizes of individual households as well as their productivity, resulting from anthropogenic factors such as poor natural resource management, overgrazing and fencing of open rangelands for private use, and reduction of mist levels are the major factors leading to the economic marginalization of the pastoral communities in the area. This also triggered the

---

[33] An evergreen shrub and a mild stimulant whose leaves are chewed mostly by men that cause euphoric effects.

influx of pastoralists to the urban area, joining those preceding them in city environments characterized by poverty, lack of space, unemployment and many other social problems, while many others turned to charcoal production for subsistence. Such scenario is a typical example of the conditions that are prevalent in all areas of the country.

3. Current and expected impacts of climate change

3.1. Precipitation and temperatures

According to community informants, the effect of the climate change that has taken place during the past few decades is manifested in the mountain landscape and its surroundings. The most visible change is the decrease of mist which used to overshadow the area for almost four months annually (from November to February). The absence of lichens (*Usnea articulata*) hanging from trees (see plate # 7), particularly the East Africa Pencil Cedar (*Juniperus procera*) indicates that that there is insufficient moisture from mist and rain to support them. Previously, the presence of mist during the long and dry winter (*Jiilaal*) season over the mountain in reasonable quantities kept the moisture regime intact, creating a conducive environment for the mountain vegetation to remain green and to thrive in that kind of weather.

According to some elderly community members, the area is also experiencing higher temperature levels compared earlier periods. There is also decrease in rainfall over the years. In a survey taken during the period from 1945-1950, John A. Hunt recorded 43.68 inches of rainfall at Go'da weyn village just at the south-eastern slope of Gacan

Libaax[34]. However, recent records for the northern highlands which Gacan Libaax is a part of, estimate the amount of rainfall at 20 inches annually. This translates a 50% decrease in precipitation levels over a 60 year period.

According to Hemming (1966), the mean annual rainfall in Gacan Libaax was only 32.5 inches; while Butzer, K. W., (1961), quoted by Hemming, has suggested that the rainfall has been reduced by 20-25% below the 1881-1910 mean value[35].

The mountain is dotted with a number of water points, most notably *Cuna Madow, Guro, Calaaculle* and *Kabcun*, which are water supply sources for the inhabitants in the area, their animals as well as the wildlife. Water levels in these water points and their recharge capacity has been decreasing to a level that it creates friction among pastoralists – particularly during the dry season. Because of water scarcity and decrease in water recharge, fetching this precious fluid for human consumption and watering animals became more arduous and time consuming as well. The more the mountain is de-vegetated, the lessor the amount of water that will infiltrate into the ground, hence, streams run dry and shallow water fall into further depths.

In the past, during the *Jiilaal*[36] season, when water is scarce, the mist descending and accumulating in rock holes and

---

[34] John A. Hunt, *The General Survey of Somaliland Protectorate, 1945-1950*
[35] C. F. Hemming, *The vegetation of the Northern Region of Somali Republic* (1966)
[36] The four Somali seasons are:
a) *Jiilaal* **(January - March)** Jiilaal is the driest and harshest season. Water and vegetation are often reduced or scarce. Pastoralists move with their animals in search of water and better grazing. If there is drought people and animals may need

crevasses provided some water for thirsty humans, livestock and wild life. It is important to note here during 1950's, when the colonial government initiated a natural resource management programme in Gacan Libaax, the few buildings established in there had roof gutters which were made into good use by harvesting the mist and rainwater falling on rooftops for consumption by the forest guards and visitors, and for raising seedling in the small nursery established in there.

---

help in accessing water. There are no agricultural activities this time of the year. Sea activities are open including fishing and boat trade.

b) Gu' (April - June) If Gu' rains are normal or above normal, they bring prosperity. Water, pasture and rangeland resources are revitalised. Milk and meat are plenty. Wedding ceremonies, regular clan assemblies and traditional dances are held during this time of the year. Animals are sacrificed to thank Allah. In pastoral tradition, a person's age is calculated on the basis of the number of *Gu'* seasons he or she has lived.

c) Xagaa (July - September) The Xagaa is the second driest season. In the coastalareas sea activities are minimised or halted due to monsoon winds. The Xagaa season is characterised by dry cool weather over most of the Somali areas except for the Northern regions where it is very hot.

d) Deyr (October - December) The Deyr is the second main rainy season. Above normal Deyr rains often lead to above average cereal production including that of cash crops and improved pastoral conditions. If both Gu' and Deyr rains fail, food insecurity or even famine could occur as the Deyr is followed by the dry *Jiilaal* season. (Source: http://www.fsausomali.org/index.php@id=77.html)

There are no more "permanent running springs in every hollow[37]".

This is a result of a combination of factors viz. decrease of precipitation, thinning of mountain vegetation, reduction in the ground infiltration of water and increased temperatures.

Gacan Libaax Mountain is an important watershed area. It is equally important for the communities living in the lowlands towards the north (coastal areas) and south where dry land farming is made possible by the seasonal water courses originating from Gacan Libaax area. Some of the seasonal flooding is used for food and fodder crop production in fertile low rainfall areas through water-spreading and water harvesting using dykes and conveyance structures. The adverse climatic conditions, particularly the decrease in the precipitation levels will cause the reduction of headwaters of those seasonal water courses originating from the mountain. A major anticipated result is water scarcity in number of villages and settlements downstream.

3.2. Food security and livelihoods

The decrease in the precipitation levels has not only affected the mountain ecosystem but also the livelihoods of the communities in the area. For example, the occurrence of the mist during the dry season kept the vegetation of the mountain green, ultimately supporting, the stockowners and their herds to endure the dry season with less hardship.

---

[37] Quoted from Ahmed Naleye, an old man who has lived on *Gacan Libaax* area all his life, by P. E. Glover in 1947

Some of the elderly community members felt quite nostalgic about the absence of the mist. Muse Aw Ahmed, a community elder said:

> "Mist is an indication of plenitude. When it was plentiful, living conditions in the area were better than today. Access to water was also easier, the vegetation was greener and more enduring, there was more moisture on the ground and erosion was not speedily progressing compared to these days. Livestock herd sizes per family were larger, more productive and with bigger body weight. The milking gourd we used in the past for milking goats is now utilized for milking camels and cattle".

In the last sentence, the interviewee's mention of the milking gourd used for goats lately being used for camels is a figurative indication in the decline of milk production levels. What he meant is that there is no need now for the big camel milking gourd anymore.

Xalwo Cige, the mother of the present writer, who has spent most of her early years in Gacan Libaax, said:

> "The grass used to hold the mist settling on it; it nourished the shoots then it seeped into the ground. The mist used to cover the area starting from *Asr* prayer time (in the afternoon) and continued to overshadow us till 9:00 a.m. in the following day. We used to call it **Hayays**, because of its incessant downpour. It dampened the soil and rejuvenated the vegetation at a time our livestock were in a dire need for pasture and browse".

Qat (Catha edulis) farming families are also experiencing decline in the quality and vigour of their plants due to the decline in precipitation levels (including mist which contributed to nourishing the shrub during the dry (*Jiilaal*) season); thus affecting the marketability of their produce.

3.3 Impact on women

The diminishing quantity and quality of animal products (ghee, milk, meat etc.) – although it affects the nutritional requirements of the pastoral household as a whole – has had its far reaching impact on women who are the ones to suffer most. These products which were, in the past, consumed by the family are now sold in the market in order to meet the shortfall of family income. Thus, women are left to eat less as food allocation among family members follows traditional gender lines which give preference to males. Data from a health centre in Go'da weyn, shows an increase of anaemic cases and malnutrition among pregnant women and children respectively. The thinning of woodlands will necessitate allocation of more time for firewood collection, acquisition of building materials for the Somali hut (*Aqal*), while water shortage will mean more walking longer distance to water points.

3.4. Biodiversity

The decrease of mist will lead to the retreat and slow disappearance of some of the forest tree species such as *Juniperus procera*. A transect walk over the mountain revealed that this species is receding to the westernmost areas and along the escarpment where the mist levels is higher but short-lived compared to earlier periods. Dead trees with shallow lateral root system, many of them poorly developed, and what looked like creepy skulls or skeletons, can be seen everywhere, particularly in the eastern part of the mountain. Some of the other plant species affected by this trend but also with poor regeneration are *Euphorbia grandis* and *Dracaena schizantha* (Dragon's Blood Tree).

With the decrease in precipitation levels, the diversity of the mountain flora has been declining over the years. The regeneration of *Juniperus procera* "Dayib" and *Euphorbia grandis* "Xasaadin" are now very low and most of the trees are showing decadence. A species of wild thyme *(Thymus vulgaris)* which used to grow in Gacan Libaax is now believed to be rare (extinct?). The community also believes that this plant thrived on the presence of mist.

The mist forest of Gacan Libaax used to hold forest animals. The decadence of the trees and opening up of the forest has already affected the natural fauna and exposed the few remaining species to extinction. Gacan Libaax is an important route for migratory birds as well as a home for wide variety of local species; therefore, its deterioration will have a negative impact on these species.

3.5. Ecotourism

Because of its exceptional natural attraction, the existence of prehistoric caves and its geographical proximity with the three major urban centres in Somaliland, there is a good potential for ecotourism, if the mountain ecosystem properly protected, preserved and kept balanced. The mountain has been an important tourist point in the past. However, the effects of climate change as well as anthropogenic agents will considerably down grade its ecotourist potentiality.

3.6. Indigenous knowledge: Traditional weather forecasting

Rain forecasting has been an advanced art among Somalis. This art was born from a synthesis of Persian, Arabian and African astronomy. For example, the *Nayruus*[38] is celebrated with ignition of fire (literally *Dabshid*) and hanging a life branch/leaf (preferably aloes) at the entrances of houses. The northern Somali forecaster is called *Xidaar* (literally: someone warning against something ominous, such as drought, tribal conflict or heralds something good such as rain) or *Xiddigiye* (an expert on weather lore). This segment of the community was very much respected as their predictions generally used to be almost precise. However, according to community informants, the predictions of these gurus are become more unreliable and their reputation is getting eroded. These gurus apply and follow the same 'formulas' and calculations which their forefathers used but paradoxically the outcome is always not as revealing as it used to be in earlier times. According to the analysis of the present writer, the disturbance in this indigenous knowledge can be attributed to the effects of climate change. This is area which requires further research.[39]

4. Current and future adaptation strategies

The main adaptation strategies adopted by the community in order to meet these challenges are:

4.1. Rural-urban migration

---

[38] Compare to *Norouz*, the Persian New Year

[39] This subject is treated in a recent research document "*The Impact of Climate Change on Pastoral Communities of Somaliland*", (November, 2009), carried out by Candlelight with financial assistance from the European Union (EU) and Heinrich Boell Foundation (HBF).

The decrease of precipitation entails reduction of the benefits to be derived by the community from the mountain resources. This leads to the reduction of families' livestock herd sizes and consequently a situation of food insecurity. As a result there is an out-migration to the urban centres. Women and elderly family members are then left to care for the livestock. In a way, this a sort of coping strategy whereby family members working in the urban areas support their families in the rural areas.

4.2. Diversification of income

In order to compensate for the loss in the livestock production system, bee keeping is becoming more common among agro-pastoralists. Poor pastoral families resorted to charcoal production, while others coupled rain-dependent farming with their pastoral production system in order to diversify their income base.

4.3. Development of new water sources/points

The decrease of water in the study area urged the economically better off households to settle in the lowlands towards the south of the mountain in order to get access to grazing areas beyond the mountain range. They construct cement lined cisterns *(berkads)* and surface water catchments for harvesting rainwater, while poor pastoral households continue their trek to the traditional wells in the mountain areas. Development of those new water sources, rather than depending on mountain water points, leads to relieve some grazing pressure from the mountain and to more even distribution of pastoralists and their animals in a wider area. It will also contribute to the lessoning the formation of livestock paths which often turn into erosive water channels.

5. Challenges

5.1. Charcoal production

Charcoal produced from the acacia trees is the main locally available energy for cooking and space heating. Gacan Libaax is located in the middle of three major urban centres in Somaliland viz. Hargeysa, Burao and Berbera with an approximate population of one million inhabitants. The area towards the south of Gacan Libaax is a semi-arid scrub which used to be well-endowed with *Acacia* species, particularly *Acacia Bussei* – the most preferred species for charcoal production. However, the thinning of these woodlands urged charcoal producers to move their commercial exploitation of trees towards Gacan Libaax Mountain. This poses a major threat to the mountain ecosystem and is a contributory factor to impacting the micro-climate of the area through increased temperature and reduced precipitation and soil moisture regime.

5.2. Thorn enclosures

The decrease of the pasture and browse condition, resulting from unsustainable resource utilization practices and compounded by the effects of recurring droughts, has triggered resource competition whereby some pastoralists enclose the communal rangelands for private use. Maintaining these enclosures require periodic cutting of trees, thus contributing to the denudation of ground cover and increase in temperature levels, soil compaction and accelerated water erosion. Moreover, enclosures are a major source of conflicts among community members.

5.3. Weak government institutions and lack of policy implementation

In Somaliland, there are a number of acts, some of them in the process of being formulated that are related to environment. The Environment policy, aimed at providing a holistic framework for environmental management is already in place. The National Environment Act, the Solid Waste Act, and Forestry and Wildlife Act are some of the legal frameworks that have been developed recently. However, it is apparent that these regulations are in a state of dormancy due to paucity of financial resources, limited technical expertise and, in some cases, their impracticality.

Community policies along these lines i.e. protection and conservation, do exist and have been successful to a certain extent. For example, the opening and closure of the grazing land of Gacan Libaax is totally in the hands of the local committee. They have also ensured that no commercial logging and charcoal production could take place in the mountain area. However, the absence of government support makes these initiatives and efforts to be potentially vulnerable to breaching.

6. Conclusion

The effect of climate change has manifested itself in the decrease of mist and rainfall, rise in temperature levels and decadence of the mist forest of Gacan Libaax Mountain. This progressive adverse ecological change can be attributed to climate change which resulted in triggering socio-economic challenges for the communities in the area. These changes have also urged them to devise adaptation measures such as shifting from pure pastoralism to agropastoralism, migrating to the urban centres and other options for diversifying their income viz. bee-keeping and charcoal production.

The local communities very well understand the environment in which they live, and have a very positive attitude towards it. They associate plenitude with the presence of mist as far as its contribution, in terms of greening of vegetation and its resulting benefits to their livestock, is concerned. Gacan Libaax is an important watershed area. A watershed can be compared to an elevated tank. There is a similarity between the way rust makes a steel water tank no longer useful and the way deforestation degrades the water storage and retention capacity of mountains. In either case the beneficiaries in the lowlands will suffer.

This is why it is important that Gacan Libaax area and other watersheds be protected and maintained.

The reduction of mist levels has had its far-reaching effects on the pastoral community in the area, causing destitution and out-migration to urban centres. Therefore, the continuation of this trend will mean more hardship, food insecurity and eruption of resource-based conflicts among community members. Other than the continuing decadence of the mist forest, some of the other areas that are sensitive to climate change are the condition of the remaining wildlife, ecotourist potentiality of the mountain and water availability.

In the absence of government support and the dormancy of the relevant policies, some of the unsound practices adopted by some community members such as charcoal production and enclosures will remain to challenge any positive action towards the mitigation of the problem.

16

Vanishing Bees

> "..And your Lord inspired to the bee, "Take for yourself among the mountains, houses, and among the trees and (in) that which they construct; Then eat from all the fruits and follow the ways of your Lord laid down {for you}." There emerges from their bellies a drink, varying in colours in which there is healing for people. Indeed in that is a sign for a people who give though". Qur'an: 16:68-69

Bees are disappearing in alarming rate across the North America and Europe. Beekeepers in the United States first sounded the alarm about disappearing bees in 2006. Seemingly healthy bees were simply abandoning their hives *en masse*, never to return. Experts are calling the worrying trend "colony collapse disorder" or CCD and they estimate that that nearly one third of all honey bee colonies in the country have vanished.

Bees play an important role in pollinating flowers and in maintaining a balanced eco-system as well as amply contributing to global food security. The list of crops that simply won't grow without honey bees is a long one and according to experts, if bees were to become extinct then humanity would perish, humanity will be next.

Why are the bees leaving? Scientists studying the disorder believe a combination of factors could be making bees sick, including pesticide exposure, invasive parasitic mites, an inadequate food supply and a new virus that targets bees' immune systems.

In the Somali region, honey is produced primarily by two races of honey bee: *Apis mellifera*, the African honeybee, which occurs between 500 and 2400 meters elevation; and, *Apis mellifera yemenitica*, a drought resistant lowland bee. Bees produce from the Somali flora some of the best honey in the world. The most popular types are acacia honey, Zizyphus honey, aloes etc. Worldwide, acacia honey is highly priced for its distinct composition, quality and taste. Somali honey is exported to Arabia and urban centres and is also used locally for medicinal and nutritive purposes.

Honey production done through gathering wild honey is important throughout the region and, in recent years, bee hiving in top bar hives is replacing the old crude and cruel way of honey collection.

Traditionally, wherever a beehive is sighted, a wild collector uses fire to drive bees away from the hives. As a result, large numbers of bees are killed. By nature, a pastoralist has no time for the domestication of bees and therefore, whenever he locates a beehive he has to make sure that he immediately harvests it, even if the honey is unripe, otherwise someone else will harvest the honey. This usually ends in the complete destruction of the hive. Centuries old "robbing" the bees of their honey and the harshness of the local environment has probably made the bees rather more aggressive.

While the bees in Europe and America are probably being decimated by pesticide exposure or something else, the numbers of local bee colonies are similarly declining due to deforestation, droughts and, of course, the cruel method of honey collection.

Nowadays, various attempts are being made by development agencies in order to introduce modern beekeeping techniques. These attempts are intended to capacitate communities in adopting and handling improved husbandry methods and giving assistance towards acquisition of hives and use of modern moveable frame hives. Beekeeping opens viable options for a sustainable and non-destructive utilization of forest resources and value-adds the remaining woodlands that are threatened by charcoal production. The good thing about beekeeping is that it has strong synergies with environmental protection and income generation. As a matter of fact, the economic value of pollination is even many times higher than that of beekeeping products like honey and wax.

17

The death of the Famous **Dheen** *Tree*

> A tree just cut by a wooden handled axe sadly remonstrated with the axe about the cruel deed: "Aah, Axe! Verily, you would not have severed my 'limps', had a section of me (the handle) was not part of you!" _____a Somali proverb

Before the year 2009, in the middle of a small seasonal watercourse, which goes by the name "Dooxa Dheenta", just by a village carrying the same name "Dheenta", on the main road linking Hargeysa to Berbera, there stood a famous tree, locally known as "Dheen" (*Berchemia discolour Hemsl.*). Then, it came to pass that many travellers passing by the tree have been kept intrigued by the unexpected drying and dying of the tree, which for many decades, if not centuries, stood majestically with its huge and all year round green canopy in contrast to the deciduous acacias in the surrounding area. They repeatedly asking themselves about the cause of its process of losing its grip on life - whether it was natural or whether its demise could be attributed to some other cause.

This tree species is becoming so rare in the country and, if someone is lucky, can only come across it along dry river beds (*Tog*). Some of the main factors that contributed to its rarity include:

- It has a tasty fruit which is appreciated by people, birds and animals, particularly monkeys. The fruit is usually consumed by birds and animals before they ripen.
- Even if seeds germinate, young seedlings are uprooted by goats.

- The leaves are palatable and are lopped by pastoralists for goats during the dry season.
- It has beautiful red wood appreciated by handcrafts people. Its hard wood is a good choice for making high sounding camel bells, preferred and liked by camel men.

The not so many living plants are all aged – an indication of its limited and slow regeneration and hence, the fame that accompanies its rarity, mainly as a landmark. This particular tree was, actually, an important landmark that is almost irreplaceable. Probably it may have been standing there for centuries, and it could have lived many more years had it not been cut short by man's action. It must have been as popular as its predecessors namely, *Xaliimaaleh, Hareeri-calaan,* and *Galool-qalinle*.[40]

The cause of its death can be directly attributed to anthropogenic factors:

- Due to its good all year round shade, truckers used to park their heavy-duty vehicles under it, to rest and sometimes perform maintenance works on their vehicles. Moreover, they used to carry out oil changes, sometimes spilling over the root areas under the tree. The oil seepage into the ground might have affected the chemistry and the intricate functioning of the tree.
- The tree trunk bore many axe marks and leaf-carrying branches were lopped as an animal feed on yearly basis.
- The establishment of a small town near the tree increased the pressure and demands on the tree. It

---

[40] *Xaliimaaleh, Hareeri-calaan,* and *Galool-qalinle*: Three particular trees believed in the past to be sacred. I narrated stories relating to these three particular trees in another book *"Dirkii Sacmaallada"*.

became a sport for children who used to climb it on regular basis, whether it bore fruits or otherwise, and hence, the constant breakage of its branches.

Prior to its complete death, I stopped by, more than ten times, in order to learn more about its condition; also perchance I may come across some seeds to propagate. On one occasion, I found with difficulty few seeds that I carried them to plant nursery in Hargeysa, run by Candlelight, in the hope of propagating them. I did some pre-treatment by boiling the seeds; unfortunately, I did not succeed that time.

The tree met its eventual fate. It became completely dry and dead. With that, the travellers resignedly stopped their appreciative glances towards the direction of the once green tree. The truckers stopped parking their heavy vehicles under it. The children and monkeys abandoned climbing it for sport and fruits collection. No more branches for the birds to perch on and build their nest on. Not to be heard are the beautiful and assorted voices of the birds, except the forlorn cry of the *fiin*. Gone also is the buzzing of the pollinators that used to crowd the flowers of the tree. Neither the camel men nor weary travellers will find leisure under its once expansive canopy, while that very place will not anymore host teenagers to play *waalo* and *Leelo goobaley*. Neither the girls, who in their reveries contemplate the lads they will join in marriage, find it a suitable place to braid hair for one another. Neither will you hear the drumbeat and synchronized clapping of the women of *Sitaad*, nor sense the aura of spirituality generated thereafter that used to surround the tree. The place will cease to serve as a venue for *Shir*[41] and *Gar* meetings. Finally, in accordance to the pre-Islamic

---

[41] Shir: *Communal and tribal meetings.* Gar: *Traditional open courts*

Cushitic belief, the spirit of *Ayaana* believed to descend on treetops in the night-time must have had bid good bye to the location.

Its fate was sealed when its skeleton was not even spared from the axe. Some residents from the village cut its dry parts into pieces, putting it into a kiln in order to produce few bags of charcoal from it. It was the very tree that gave the name to the village they reside. Now Tuulada Dheentu stands there without an identity, and an important piece of its history lost forever.

18

Toxins from Tanneries and their Effects on Environment and Human Health

> "Only when the last tree has died and the last river have been poisoned and the last fish been caught will we realize we cannot eat money."____ Cree Indian Proverb

For many centuries and more, the Somali inhabited areas of the Horn have been well endowed with livestock (sheep, goats, camels and cattle). Animals for export to the traditional Middle East markets were shipped on the hoof. It is only in 2000 when slaughterhouses in different parts of region started shipping carcasses by air. In the olden days, unprocessed hides and skins were also major export items. There were no tanneries in the country, although hides and skins were treated with pesticide. It was only in the early 1970's when the Somali Government established the Hides and Skins Agency with the aim of developing this sector as well as value adding these products through the establishment of a number of tanneries in the country as well as disseminating better techniques in skinning and drying these items. In Somaliland, two major tanneries, one in Hargeysa and the other in Burco, were positioned in secluded locations outside the towns in order to decrease the risk of pollution. However, after forty years, both sites are now engulfed from all sides by buildings. The Hargeysa Tannery was completely destroyed during the war of 1988-1990. The debris was dismantled and the site is now converted into a residential place. The Burao Tannery is still intact with occasional tanning processes undertaken from time to time – even though internally displaced persons (IDPs) have encroached on its premises from all directions. This was not the end of tanning activities in the country. There are a number of smaller enterprises operating in the residential areas of major towns, Hargeysa in particular that carry out some of the processing stages of hides and skins before they are exported to their foreign destinations.

The high demand for leather products on one hand, and the increasing environmental restrictions on tanneries in the developed countries, whereby substantial capital investment are required for effluent treatment facilities on the other, are expected to push more smaller tanneries to move to the poor Third World countries like Somaliland, where labour is cheap and environmental policy restrictions are non-existent.

The town of Dacar-Budhuq, which is situates on the Hargeysa-Berbera road, and at the same time the administrative seat of Laas Geel District,[42] is the site of two major tanneries.

The two factories were positioned on each side of the seasonal watercourse that brings vital water for Dacar-Budhuq community as well as many settlements downstream. The water course empties its seasonal water into the Gulf of Aden. Both factories are on the upstream side of the town and are not very far from the village water source (actually one of them is just a stone's throw from the only protected well for town water supply). Both tanneries are on the sloppy banks of the watercourse that could increase the risk of water contamination. Dacar

---

[42] *Laas Geel district: named after the site of the ancient cave paintings, situated 50km northeast of Hargeysa)*

Budhug has the potential of becoming a major town. The population of the town dramatically increases during the summer as many families in Berbera leave behind the unbearable hot weather of Berbera.

These recent developments in Dacar-Budhuq are now a major subject of discussion and a good cause for worry for the residents of the town. While many of them see this as a golden opportunity that could lead to employment creation and overall development of their town, others are worried about the pollution problems caused by tannery waste that can downgrade the quality of ground water, and thus impacting the residents' wellbeing and affecting livestock health as well as the area's agricultural potential.

Generally, tanneries discharge waste. The effluents when discharged to neighbouring surface water deteriorate the physical, chemical, and biological properties of the receiving body (water). Organic matter loads lead to rise of noxious odours and to depletion of the dissolved oxygen in water. Suspended solids cause turbidity and when settling onto the bottom of the water destroys living life. Chemicals and toxic residues render the water unsafe for any domestic use and thus represent potential hazards to human health. Respiratory diseases and skins infections are known to be common among tannery workers.

Worldwide, international consumers and national governments are increasingly applying ever-tighter demands on the tanning industry to treat wastewater and solid wastes according to international standards.

Consequently, industries face raising environmental costs to achieve increasingly tightening discharge consents. This could be the reason why some industries discharge their industrial waste in a manner that contravenes international regulations. In our case, it is apparent that neither the high risks of waste from these tanneries have been discussed, nor Environmental Impact Assessments (EIAs) was carried out. Our institutions lack the expertise and the political will to deal with issues like this one. However, this gap could have been bridged by commissioning environmental engineering firms that could have issued an Environmental Health Check at the cost of these companies.

After many years of operation, the effects of the effluents on human and other life forms are surfacing. Skin and respiratory diseases such as eczema and allergic asthma are few of the potential health risks. The highly toxic chromium salts, acids, and ammonium salts used in huge quantities can cause huge health and environmental impacts. Infertility and low birth rates of humans and livestock could be another impact.

It is never too late! The situation calls for a quick action aimed at controlling the imminent pollution problem that could result from these as well as future tanneries. The risk can be greatly minimized by establishing effluent treatment plants and a strict monitoring programme. It is also very crucial to ensure that tannery workers are provided with safety gear such as coveralls, gloves, safety shoes, respirators and eye protection materials.

Ye-eb (Cordeauxia edulis)

> "A soup prepared from Yicib (which is not only nutritious but delicious as well) is hard to come by"___ Somali proverb

Ye-eb (*Yicib*) is an evergreen shrub native only to Central Somalia and areas in the Somali inhabited region of Ethiopia bordering Somalia. The shrub which is called *Gud (Cordeauxia edulis)* produces nutritious nuts, also called "*Yicib*" which can be eaten directly, roasted or boiled in water. Yicib nuts are highly nutritious and are tasty with a chestnut-like flavour. When the fruits are fresh they are locally known as "*balag*". Once they are separated from branches, they are deshelled, spread and dried in the open. Such dry nuts are called "*Kalmoon*". Then, they are roasted. If the fruits dry on the branches of the shrub, they are known as "*jalow*". Roasting gives the fruits a pleasant smell and better digestion. *Yicib* nuts are considered as one of the best gifts to carry from the area where it grows. Seeds germinate easily, but early top growth is slow, as the plant's top priority is to develop massive root system to draw water it needs for its survival in the harsh environment.

It has long been said "a soup prepared from Yicib (which is not only nutritious but delicious as well) is hard to come by" as different users namely humans, livestock and the squirrel (*dabagaalle*) compete for its fruit. The nuts are much relished, often being preferred to usual diet of dates and rice.

Two variety of Yeheb are recognized; *Suuley*, a bigger variety and can only be found in central Somalia, i.e. south of Mudug, Galgudud and north of Hiran regions. *Muqlay*, a smaller variety which is found in Central Somalia, Somaliland and the Somali region of Ethiopia, as well as in central Somalia.[43]

The leaves of the plant can be used as tea-leaves, and are also browsed by sheep, goats, camels and cattle. The bones of animals feed on the foliage of Yicib become stained with a bright orange colour. This is due to a brilliant red natural dye called *Cordeauxiaquinone* can be extracted from its leaves. This is a unique and insoluble dye, used by dying factories. The dried twigs of the plant make one of the best firewood in the country.

Yicib has been tested in countries outside Somalia, such as Israel, Kenya, Tanzania, Sudan, Yemen and the USA. Before the civil war in Somalia, fifty shrubs were cultivated in the Central Agricultural Research Station in Afgooye. Plants grew slowly, but flowered and provided fruits abundantly.

The plant is always free of pest, but the nuts are attacked by weevils and moth larva. The production capacity of a mature Yicib plant is 5 kg per year; however, a very small percentage can be harvested as nuts are eaten by the non-human competitors even before they reach maturity. Moreover, the indiscriminate cutting of Yicib sticks for

---

[43] Dr. Muna Ismail, Lewis Wallis and Scot Draby, *Restoring Land and Lives: Report of a scoping mission to examine the restoration and possible domestication of the Yeheb plant in Somaliland*, (June 2015), Initiatives of Change, London, UK.

shelter qualifies it as an endangered species. This fact was recognized by the previous Somali regime and, therefore, counted it as one of those plants to be protected from cutting and/or burning. During 1980's, an area measuring 25 hectares at *Salah-dhadhaab*, between Belet-weine and *Dhuusa-ma-reeb* was fenced with the intention of conserving the shrub in its habitat.

Generally, the plant is highly resistant to droughts in semi-arid climates. For example, during the 1973-76 drought, the Yicib bushes survived, although it failed to flower and set fruits.

After many years of civil strife and absence of functioning government in Somalia, Yicib is in a critical situation and may face extinction due to over-utilization and absence of conservation measures. There is a need to ensure that the plant is not banished from its habitat, and to sustain its multi-faceted benefits to humans and animals. Therefore, I would encourage development agencies (local non-governmental agencies in particular) operating in Hiran and Mudug regions of Somalia, as well as those in Region V of Ethiopia to fence, at least, a small area of Yicib stand so that its natural succession is maintained, even in such confined locations. Domestication of Yicib is another option. Tried elsewhere, its introduction will not only contribute to its existence, but will also create new source of income and nutrition.

20

The Pillage and Plunder of Somali Marine Resources

"Among all His servants God shares out the bread;
Whether it is the fish in the sea or a cup of tea;
Every person shall receive what it is allotted to him;
Even though he sets out in the morning, or runs or climbs a high hill;
That no man will gain more (than his share), let that be known" *(From a poem by Ismail Mire)*

Although I tried to capture and sum up in this book the different aspects of the Somali environment, the discerned reader could easily notice that the author is biased towards terrestrial environment, and the mention of the marine resources is very limited. This does not mean that the sea is of less importance to people than land. The basis of my choice, however, was the fact that the most ideal habitat of humankind is the Land and, therefore, their behaviour on the land could also influence or determine their exploitations in the sea. For example, if we decide to protect and conserve our land-based natural resources and life forms, undoubtedly we will also act responsibly and take a reasonable care of our marine resources.

Culturally and historically, Somalis have never been dependant on fish for their nutritional needs and their knowledge of the sea was very limited. This could be the reason why Somalis called the "man-eating fish" as "Sea lion" {*Libaax-badeed*}. The same as they regarded the lion to be the "king of the jungle", they also thought shark to be the "king of the sea".

The Somali territory the longest coastline in Africa (c 3,300 km) and one of the five most productive upwelling regions in the world that has significant potential for offshore fisheries development. The prolific fisheries can be associated with the upwelling of the Somali Current off the north-eastern coast of Somalia.

As land-based resources continue to be degraded and over-exploited on one hand, and human needs similarly continue to increase on the other, more people are venturing to earn a living from the sea. The former Somali government recognized the importance of introducing fish to the population as a major source of protein. The government operated a fishing fleet that employed over 30,000 people, and contributed to 2% of the gross national product (GNP)[44]. With the collapse of the Somali government, the entire fishing infrastructures such as cold storage, maintenance workshops, and most importantly the Las Qorey fish factory were either looted or fell into state of disrepair resulting in the loss of livelihood for thousands of people who used to eke a meagre existence from the fishing sector. The sea has become 'free- for- all'. Anyone can fish wherever and whenever he desires and there is no time frame or seasonal calendar that allows people to fish in certain periods and not to go to sea in the spawning periods of the resource. However, the most serious problem constraining sustainable harvesting of marine resources is lack of investment and growing evidence of significant 'poaching' by foreign commercial fishing fleets inside the 12 nautical mile territorial range of

---

[44] Yearly Fisheries & Marine Transport Report, 1987/88, Ministry of Fisheries & Marine Transport, Somalia Republic.

the coast exploited by artisanal fisheries. Many of these fishing fleets use illegal, unreported and unregulated (IUUs) fishing practices including the indiscriminate use of all prohibited methods of fishing such as drift nets and under water explosives on fishing breeding grounds with the aim of maximizing their fish catches. Such practices would lead to the acute decline of fish species, loss of biodiversity and ultimately endanger the means of livelihoods of both artisanal fisheries as well as the quality of life of the Somali people. Selective fishing practices whereby the high value species caught in the nets are collected and the rest dumped into the sea is an irresponsible act as much as it is waste of resources. Among the many conservation measures adopted in sustainable fishing, a minimum mesh size to let small fish escape to replenish stocks should have been adopted by fishing fleets. Unfortunately, even if there are policies set to address best fishing practices, local institutions do not have the capacity to implement such regulatory instruments. Hence, the quick depletion of stocks being experienced in the Somali waters.

The adoption of fishing regulatory policies by many nations in response to the declining fishing stocks is another factor driving many international fishing companies to fish in the Somali waters. For example, the Common Fisheries Policy of the European Union member states sets quotas for which member states are allowed to catch what amounts of each type of fish.

As international demand for nutritious marine products increases and as the fear of worldwide food shortage grows, uncontrolled and unprotected Somali seas became the target of the fishing fleets of many nations[45]. Dumping of industrial, toxic and nuclear waste in both off-shore and onshore areas have been reported repeatedly by Somali authorities, local fishermen, civil society organizations and international organizations and warned of the dangerous consequences of these criminal actions.

---

[45] From an article by Mohamed Abshir Waldo "*The Rape of Somalia by Illegal Fisheries*".

21

The Link between Droughts and Deforestation

Oh Bullo!
Even the Oryx made a long search, for a place with good grazing, not far from water point, but failed to find it.[46]
_____Traditional Somali pastoralist work song (in the form of conversation with livestock) and a morale booster sung when moving animals during their trek to a pastureland far from water points.

Drought-related emergencies in the Somali region have almost become yearly occurrences and as communities' coping mechanisms become weaker, effects of droughts are equally becoming more severe and hard-hitting. Alerts and alarms are signalled by Government authorities as well as agencies specialized in emergency preparedness and response; then resources are mobilized and spilled over the affected communities, yet again the NEXT drought hits in the same unprepared and disorganized situation as in the earlier emergency. Emergency intervention (dubbed as life-saving), water trucking in particular, often lack linkage with the structural causes of drought situations, and have become a costly yearly ritual. As a result, many supporting agencies are becoming more unwilling to positively respond to such situations.

---

[46] Overgrazing is generally normal near water points due to the constant presence of grazers and browsers, so herders move their stock to distant areas.

I think it is high time that development agencies and government planners should discuss the root causes of drought in an arid environment such as ours. Drought is defined as: *"an extended period of months or years when a region notes a deficiency in its water supply. Generally, this occurs when a region receives consistently below average precipitation."*[47]

The Somali-inhabited areas and the Horn of Africa in general are drought-prone and one of the most important factors leading to those drought conditions, other than climatic factors, is the massive destruction of trees. Charcoal production, wood for construction, and establishment of thorn enclosures are some of the main activities leading to deforestation.

Charcoal production is causing the impoverishment of rural communities, pastoralists in particular, as the rangelands are denuded of vegetation cover and livestock is decimated by shortage of grazing and consequently succumb to animal diseases. Pastoralists feel helpless and powerless against the exploitation of charcoal traders who provide food and *Qat* to their youth and exchange with charcoal.

Denuding the country's rangelands leads to a number of severe environmental consequences including:

- Loss of forage (trees, shrubs and herbaceous plants).
- Loss of water retention in soils: The denuded ground is exposed to the direct hit of raindrops that create a hardpan (a hard, compacted, often clay layer of soil through which roots cannot grow).
- Loss of biodiversity.
- Unfavourable changes in micro-climates, particularly temperatures, rainfall and wind.

---

[47] R. N. Sahay, *Ecosystems and Water Resource Management*.

- Loss of carbon dioxide sink thus more gas in the atmosphere that leads to global warming and its effects;
- Increased sheet and gulley erosion;

In the light of the above, addressing the problem of deforestation can directly contribute to mitigating the effects of drought.

Somalis generally give a name to every serious drought occurring in the country. For example, the most popular name among pastoralists, given to the drought of 2009 was *Garowle* (literally the Year of the Sorghum). It was named so, because for the first time in the history of Somalis, sorghum became an important feed for livestock. The deteriorated pasture urged pastoralists, who were already burdened by the cost of trucking water for their animal and household consumption, to purchase sorghum, not only for human consumption, but also as an animal feed. That was an indication of a serious grazing shortage.

Given the fact that the domestic economic activity is dominated by livestock production, and the core Somali cultural is oriented around the pastoral way of life, this unabated destruction of the remaining woodlands is tantamount to a direct and long-term assault on the Somali people.

This calls for a strong action by concerned government officials and departments to address this issue of deforestation if we have to be serious about addressing recurring droughts. Equally, development agencies should not always wait for the next drought to occur, but extend their crucial support in restoring rangelands and address the root causes of droughts so as to contribute to the reduction of the vulnerability of rural communities.

22

Honey Adulteration and Testing its Purity

> Consider the aloe - how bitter is its taste!
> Yet sometimes there wells up a sap so sweet
> That it seems like honey in your mouth.
> __From the poem: *'Sweet and Bitter'* by Ahmed Ismail (Qasim)

The dramatic decrease of bees across United States and Europe, caused by the worrying trend which experts call 'Colony Collapse Disorder' on one hand, and the resulting increase of honey price on the other, and the limited supply of honey worldwide is tempting some bee keepers and food processors to sell impure honey containing inexpensive sweeteners such as corn syrup and other ingredients. Some adulterated honeys, are almost indistinguishable, physically and chemically from real honey and is, therefore, finding its way into markets causing health problems to consumers.

In 1997, a contagious bacterial epidemic have slashed down honey production in China and the beekeepers, thereafter, who had only two choices, namely, either to destroy the hives or to apply antibiotics finally opted for the use of chloramphenicol to kill the bacteria. Chloramphenicol is so toxic, only used in life-threatening infections in humans, and then when all alternatives are exhausted. Honey tainted with this drug was detected in the European and United States markets causing thereafter the ban on the importation of Chinese honey in these countries. Unfortunately, this did not stop adulterated honey from entering these markets as well as others in the third world. Honey laundering is now a widespread business practice. This is an unethical trading technique whereby adulterated honey passes through another country before it reaches the destination country. These 'honey' is then re-processed, labelled anew as a product of the intermediate country.

By doing so, the consumer is made to believe that the honey is of high quality.

Now the question is how can someone tell whether the honey he/she is buying is pure?

Some legends say:

a) If you dip a matchstick into honey, it will still light up when you scratch it against the side of the matchbox.

b) If the honey you put onto tissue paper does not wet it, it is pure.

c) If you put a drop of honey on a sandy ground, pure honey will not seep into the ground but stay protruded in a semi ball shaped structure. Contrarily, if it is adulterated or fake honey, parts of it will seep into the ground

These three legends are just tests for moisture content are not fool proof. You can get the same results from motor oil or glucose/fructose syrup.

There are only two ways that you can tell if honey is pure:

- The first is if you have a lab or access to a laboratory.

- The second is, keep bees for a couple of years. Then you will learn the taste of pure honey, depending on where the bees have collected the nectar they ripened into honey.

There is no other way you can judge if honey is pure.

Another question related to the purity of honey is: If honey crystallizes, is it pure?

All honey "granulates" or crystallizes in time, more so, unheated and unfiltered honey; some just crystallize sooner than the others. Granulation is when the bottom or all of the honey in a bottle turns semi solid and the colour turns into a creamlike shade. When you try to scoop it out, it will exhibit a consistency like peanut butter only a little grainier. Honey crystallizes because it is a supersaturated solution. This supersaturated state occurs because there is so much sugar (dextrose) in honey (more than 70%) relative to the water content (often less than 20%). This happens because the honey has more dextrose than fructose and the dextrose tends to 'go out of solution' and granulate. The added bonus for granulated honey is that you can spread it on toast, bread etc. Most Westerners prefer granulated honey. Other factors that play a role in the crystallization of honey are the temperature of where it is stored, and the type of container it is stored. For example, honey stored in plastic container crystallizes faster than another put in a glass jar.

23

Me and My Toothbrush Tree

> If it were not too much a burden on the believers, I would prescribe that they use the *Caday* before each prayer" ____ *Hadith* by Prophet Muhammad (PBUH)

Salvadora persica is an evergreen shrub or a small tree reaching 6-7m. Its habitat is widespread and mostly found in the drier parts of

Africa, the Middle East and the Indian Subcontinent and best adapted to alkaline or very saline soils. Its leaves are oblong elliptic, dark green, rather fleshy (succulent) with high water content (15 to 35%) and are rich in minerals. The fruit is spherical, fleshy, 5-10 mm in diameter, pink to scarlet when mature and single seeded. The trunk is usually crooked, seldom more than one 35 cm in diameter, bark scabrous (rough to touch), cracked and whitish in colour. The generic name, *Salvadora persica*, was given in 1749 in honour of an apothecary of Barcelona, Juan Salvador y Bosca (1598-1681), by Dr. Laurent Gacin, botanist, traveller and plant collector. The true specimen of this species came, as the specific name indicates, from Persia.

The most popular use of *Caday* is tooth-brushing. It is believed to remove tartar and refreshes breath. Leaves make good fodder and are readily consumed by goats and camels and are said to increase milk production. The high salt content of the leaves is said to affect the taste of the milk. The plant is a good source of nectar and its wood is sometimes used for firewood and charcoal. However, it is not used for cooking meat, as it leaves a foul taste. Grown in plantations or hedges, the tree has potential for reclaiming saline soils. Coppices well and branches are cut repeatedly to produce short stems that are harvested for toothbrushes.

Ask any Somali, which plant does he/she favour most among tree species, and the reply is very likely to be "*Caday*" (pronounced as Adhai'). While toothbrush twigs can be cut from many plants, this tree is second to none; and because of its oral hygiene use it is sometimes dubbed by the local people as "*Af-wanaaje*" (literally: the one that makes the mouth good and clean). Twigs cut from small branches and roots, 3-5cm in diameter, are used for toothbrushing. *Caday* is the "companion" of every Somali, the devout people in particular. One can see it inserted in the breast pocket or held in the mouth, and in the olden days, before the arrival of European clothes with pockets, one could always see it tucked behind the ear. Giving a person a *Caday* twig is regarded as a gesture of friendliness and generosity that no giver expects that the receiver would refuse the offer. Compare this to a westerner who offers a toothbrush to some stranger, where the later would regard the act as an insult! Even nowadays, at a time when local markets are packed with many brands of toothpastes and brushes, the widespread utilization of *Caday* has not been affected. A Somali travelling to any destination outside his/her homeland often carries with him/her a small bundle of *Caday*, as a valuable gift to friends or relatives living outside the country.

Personally, it has been one of my dreams to germinate this wonderful plant and then distribute the seedlings to others so that they raise them in their homes, in the hope that I will be rewarded by Allah for such a good deed. I started my trials back in 2002; however, it took me five years before I managed to get hold of few seeds, and I then 'caused' them to germinate inside the compound of my house in Hargeysa.

Prior to this seemingly notable success in the propagation of *Caday*, for a number of years, whenever I go to the countryside and thereafter come across a *Caday* shrub, particularly during summer time when they are likely to bring forth their fruits, I used to stop by it, perchance I may find some of its seeds. Unfortunately, I had to leave it empty-handed.

For all those years that I had no luck in coming across *Caday* seeds, I have been intrigued by the 'absence' of seeds from the trees. I even heard some people mention that it has no seeds at all which I did not agree with. Sometimes I had to crawl on all four under the low-lying branches in order to get hold of the tiny seeds.

Then, in one summer morning, at the beginning of June 2009, an unexpected thing that unravelled the puzzle of the missing *Caday* seeds occurred. One of the seedlings which I planted inside the compound of my house brought forth its first seeds after the flowering phase. I have been waiting with little patience for the seeds to mature, and at the same time, I used to monitor their growth on daily basis. Then one fine morning, after I performed my morning prayers, while I was still in my bedroom, I heard voices of tiny birds – the type which Somalis call "*Yaryaro*" (literally smallish). They have high pitched voices compared to their small bodies... *jiiq jiiq... jiiq*. The first thought that came into my mind was – in line with the traditional belief – whether they were harbingers of ominous thing to happen, such as the presence of a snake in the compound. Coming out of my room, and

Behold! A flock not less than thirty birds were perched in the branches of my *Caday* tree, devouring and relishing my Toothbrush tree fruits and seeds.

The sight was amusing and, I looked helplessly at the scene with a gaping mouth, as those tiny creatures carried away my seeds.

On a positive note, I was reassured by the noble biological function these birds were fulfilling: After they eat the fruits and the seeds pass through their guts, they poop them up. The process loosens the endocarp, then the seeds germinate easily – bearing in mind that the fruit pulp of *Caday* seeds contains germination inhibitors which can be removed by the digestive system of birds.

The sight of birds perched in the branches of my Caday shrub reminded me the parable of the Mustard tree in the Bible[48] which some botanists refer it to the Caday tree or is that they were amid their morning "prayers" and glorification". The verse in the Qur'an that flashed on my head was:

> "Don't you see that it's God whose praises all beings in the heaven and on the earth do celebrate and the birds (of the air) with wings outspread? Each one knows its own (mode of) prayer and praise. And God knows well all that they do". (Qur'an 24:41-42).

---

[48] "The kingdom of heaven is like a mustard seed which a man took and planted in his field. Though it is the smallest of all your seeds, yet when it grows, it is the largest among herbs and becomes a tree, so that the birds of the air come and perch in its branches." Matthew 13:31-32.

Finally, having deciphered the puzzle of missing seeds, there was nothing else left for me except to design my strategy aimed at outwitting the birds in order to harvest the seeds: I simply bought from the market a mosquito mesh wire and cut it into pieces enough to tie them around a number of fruit bearing branches to protect them from birds. This was how I managed to harvest a good stock of Caday seeds to fulfil my cherished dream of propagating and distributing them to others.

24

Rapshie Island

> "There is not an animal (that lives) on the Earth,
> Nor a being that flies on its wings,
> But (forms part of) communities like you"
> ____ The Holy Quran: 6:38

Maydh Island (*Buurta Rabshiga*) is the name given to that magical place located c. 13 km offshore, to the northeast of the coastal town of Maydh in Sanaag region. It is little over 1.5 km in length with a maximum width of 300 m and an average height of 100 m. The long axis of the island is oriented from east-north-east to west-south-west.

The Island is an important breeding place of sea birds. In 1946 the number of birds present, as report by North, M.E.W, was estimated at c. 100,000, but sadly were reported then to be declining. The Island is devoid of vegetation but rats and snakes are reported to be present there.

The following account was narrated by Abdi Ali Jama, head of Nature Somaliland, an environmental non-government organization focusing on ecotourism, research and conservation of birds. Jama had visited the Island in 2008 on an assignment to survey the coastline of Somaliland for birds. I requested from Jama to write the following essay, as I knew no one else can give justice to such a unique topic.

"We were offered a unique project to survey the coastline of Somaliland for birds by Wetlands International in January, 2008. We took a boat from Berbera and it took us five days to get to Maydh Island. The excitement and anticipation we had nursed during and prior the few days was rewarded immensely by the sheer beauty and utter unworldliness of Maydh Island, Rapshie or just the great 'Rock' as I began calling it.

Geologically, the site is not really an island. It's merely an awe-inspiring extrusion of a granitic rock from the sea-bed with no conventional vegetation whatsoever. Not an island so to speak of; just one dark-grey natural obelisk; extremely steep on all sides. The north-east side is broken up a bit into four mammoth parts by the eternal effects of the harsh weather. The south-west side has a small field of scree and single-storey-size boulders. A small group of intrepid souls can pitch a few one-person tents on this field. All other the sides are very steep and the entire rock is covered in guano from largely breeding sea-birds; mainly two species of gulls and an excellent variety of terns and boobies. It's quite a sight to behold.

In our January visit, the 'Rock' was by and large empty of gulls and terns but teaming with three species of boobies, Red-Footed, Masked and Brown. Their unrelenting noise and the prevalent stench of the guano overloaded our senses. Unfortunately, the boat captain refused to try to land there so we could hop about in the scree field. Understandably, he was afraid of ramming a submerged rock or getting tossed against the steep sides of the giant 'Rock'.

The 'Rock' is really of little interest to Somalis and very, very few visitors make it there whether scientists or other types seeking to see this natural wonder.

While we were there, the 'Rock' was surrounded by at least thirty Yemeni boats heartily fishing the fill of their huge nets. We warned some of them about scampering up the scree field. They had ropes and big axe-like machettes. We thought they were in the process of getting ready to scour the steep 'Rock' walls for guano[49]. There were no law enforcement personnel or any other Somalilander presence there.

The time to visit this magical site is between mid-October and mid-June. From mid-June until late September, when the 'Rock' is rocking with breeding gulls and terns, the sea happens to be much too violent due to the *Xagaa* winds and anything smaller than a large skiff won't dare take you there. We tried to return in mid-August of last year but found the sea much too violent and even Maydh village was abandoned. This is such a neat natural cycle that by the time the winds die down, the sea settles down and the Yemeni fishermen return, the magic vanishes and the gulls and their new broods disperse out to the shoreline. Contrary to popular thinking, especially in Somali folklore, the breeding bird life never migrates to Europe or other far northern points as done by other migratory species. Some may venture to the northern side of the Gulf of Aden, in Yemeni or Omani shores but that is about it. The breeding gulls are *Sooty, L. hemprichii,* and the slightly similar *White-*

---

[49] Guano: Is the excrement (feces and urine) of sea birds bats etc. Guano manure is an effective fertilizer and gunpowder ingredient due to its high levels of phosphorus and nitrogen. The material is mostly transported to the port of Mukalla in Hadramout district of Yeman as fertilizer in tobacco production.

Eyed, L. leucophthalmus. The largest tern species breeding there is the Greater Crested Tern, S. bergii.

Maydh Island, Rapshie, or merely the great 'Rock' is one of the six Somaliland Important Bird Areas (IBA's) as such designated by Birdlife International. It should be a Ramsar[50] Site but that is far, far away right now. I am determined to see this site during the breeding season at least once before I move on to the other world".

---

[50] This relates to the Ramsar Convention (formally, the Convention on Wetlands of International Importance, especially as Waterfowl Habitat) which is an international treaty for the conservation and sustainable utilization of wetlands, recognizing the fundamental ecological functions of wetlands and their economic, cultural, scientific, and recreational value. It is named after the city of Ramsar in Iran, where the Convention was signed in 1971. (Source: Wikepedia)

25

Essay on Environment

{Arlooy! faro kaa badan iyo biyo kaa badaniba wey ku hafiyaan"
Oh "Mother" Earth! "Both over-population and deluge overpowers and inundates you" __ Somali proverb

The word "environment" means the external conditions affecting the growth of plants and animals. It also means the sum of all living and non-living things that surround an organism or a group of organisms. In addition, it is the circumstances that surround a biome. An environmentalist is a person who is concerned with, or advocates for the protection of environment. Similarly, environmentalism is a political and social movement with the aim of protecting the environment through a combination of various actions and polices oriented to nature preservation. In the absence of clean and protected environment, the whole existence of human beings, as well as other living things is at the great risk of extinction. The responsibility of protecting this fragile environment on earth lies on the shoulders of Man. In other words, he is God's vicegerent on earth.[51] Therefore, we have to ensure that the ecosystems function in harmony to maintain *the* equilibrium and the health of the planet and of all living organisms.

Yet our actions are leading to destruction rather than sustaining life and ensuring its continuity for meeting the

---

[51] Holy Qur'an 2:30

needs of the present and without compromising the needs of the future generations.

The biggest environmental challenge facing the planet earth is pollution. Pollution is contamination of the natural environment with harmful substances often as a consequence of human activities. There are a number of different types of pollution that have far-reaching negative impacts on Earth. These include land contamination, water and air pollution. The use of fossil fuels is the biggest contributor to the buildup of carbon dioxide in the atmosphere, which leads to greenhouse effect and global warming. Some of the other important air pollutants are sulphur dioxide, nitrogen oxide, volatile hydrocarbons, carbon monoxide, methane, noise and Chlorofluorocarbons (CFCs). Pollutants that affect fresh water are sewage and inadequate sanitation, fertilizers, silt and toxic metals. Pollutants affecting marine life include sewage and inadequate sanitation, fertilizers, oil spills and plastics. Pollution not only affects humans, but also flora and fauna. For example, the depletion of the ozone layer leads to increase in ultraviolent rays which slow down photosynthesis and growth in phytoplankton. As phytoplankton absorbs carbon dioxide and releases oxygen, its decline will contribute to the greenhouse effect. Turtles mistake plastics for jellyfish resulting in breathing and digestive problems and ultimately to their death. Oil slicks, chemical effluents and sewage litter the land and seas and cause problem for living organisms. Also evidence of increasing air pollution manifests itself in the form of lung cancer, asthma, allergies, and various breathing problems along with severe and irreparable damage to flora and fauna.

Current conservation efforts towards the protection of environment include utilization of new resource-efficient technologies, although still in their infancy stage, which are aimed at reducing world consumption of fossil fuels one hand and contributing to a cleaner environment through low carbon emissions, on the other. In the recent past, more focus has been directed towards the sustainable use of water, land and other scarce resources, recycling and reduction of waste, preservation of endangered species and protection of biodiversity. The concern for environmental protection has also triggered political and social movements who advocate "green living" and the use of "earth friendly" goods as well as promoting environmental considerations in decision-making. National and multilateral environmental agreements on environmental protection have been signed although there are gaps in their compliance and enforcement.

Along with the above activities, there is a need to intensify research and use of renewable energies (solar, wind, tides, rain, and geothermal heat) and reduction of dependence on fossil fuels. It is also worth noting the relationship between population explosion and environment. The earth can be compared to a spaceship carrying its crew. The resources (food, water air) in it are limited, finite and subject to pollution while the crew is growing larger day after day. Live on earth began 4 billion years ago, and the current population is over 7 billion. The negative consequences of this rapid population growth are many, such as loss of biodiversity, hunger and famines, conflicts and wars. Despite of these challenges, there is an ever increasing re-awakening of world nations to tackle the environmental problems facing our planet and our very existence.

There is this growing trend towards attitudinal change by leaders and educated people. Environmental issues are now high in the agendas of many governments. At least, it is worthy here to quote President Barack Obama, during his inaugural speech, to affirm his commitment for cleaner world and putting the care for environment as a fundamental priority issue. He said:

> "We will harness the sun and the winds and the soil to fuel our cars and run our factories….. With old friends and former foes, we will work tirelessly to lessen the nuclear threat, and roll back the spectre of a warming planet".

We hope statements of the likes of the above, by world leaders, will not be only for political campaigning to win more votes, but precipitate to action.

26

Where Have the Birds Gone?

Year after a year, the sky above us is becoming devoid of birds. During decades gone by, the sky used to be crowded by vast numbers of birds flying leisurely in flocks. Seeing them heralded freedom; their sight personified beauty, while their comings and goings was equated to news bearers (be it positive or negative). For thousands of years, birds used to capture the imagination of humans, nurture it, and, at the same time, afforded it with wings to carry man's reverie and thoughts to distant planets and far beyond.

Even in the recorded human history, there are people who attempted to fly like the birds, whether it was for the sake of sport, or adventure, or even because they wanted to achieve surreal and internal independence. Man's scrutiny on the birds' hydrodynamics urged some people to attach their bodies with feathery wings. Still, such attempts usually ended in failures and mortalities. A famous and a celebrated incident of such an attempt appeared in an ancient Greek mythology on an old man and his son. As can be gleaned from the story, it had a disastrous end as the young man died after he belittled his father's advice:

> "Remember to fly midway, for if you dip too low,
> The waves will weight your wings with thick saltwater,
> And if you fly too high, the flames of heaven will burn them from your sides.

Then take your flight between the two. "[52]

The wisdom in this story is reflected in the Somali proverb "a young man misses or ignores an advice as much as he can jump over a tree" (***Nin yari inta uu geed ka boodo ayuu talo ka boodaa***). The cause of his demise is that he flew far up into the sky, as he was so happy to do so, and could hardly believe he can fly as bird, as well. Naturally, his soaring height took him far nearer to the son to melt the wax they used to attach the winds on his arms, to which head fell on the ground.

In Somali culture, sighting, companionship, and hearing or listening to the voices of the different birds, had different meaning for the people; and the lack of verbal communication between the two was not a barrier at all to bar them from interacting and communicating with birds, sharing with them their feelings and concerns, prophesying on the future, and sometimes using them as 'messengers'. Worldwide, stories of birds trained to carry written messages between leaders in times of war and peace have been reported, or more inexplicably, if not miraculously, Prophet Suleiman who was endowed with the power of understanding living things used the hoopoe as a message bearer.

The Somali word for bird is '*shimbir*'. The word is also interpreted as 'a new lease of life', 'recovery from a terminal disease' or a 'harbinger or a medium that brings good tidings relating to the recovery of a sick person'. For example if a person is terminally ill and nothing supporting his recovery can be arranged for him, Somalis say '*hebel shimbir ma laha*' (literally, that man has/she no bird' which means he 'has no luck to continue living', but

---

[52] Daedalus and Icarus, from Ovid's Metamorphoses, Book VIII

in case he/she miraculously recovers from illness with a potion concocted by someone and then later recovers, people say *'Shimbir baa hebel waxaas looga soo dhigay'*, {such a person or thing was a *shimbir* for the sick person to recover}. It is worth noting here that the Somali term *'shimbir'* is meant for both male and female bird, but with slight difference in the pronunciation. On the negative side, and according to an old popular belief, *shimbir* infliction is when a person experiences temporary weakness or paralysis on one side of the face. Modern medicine, however, has proved that this could be due to ill-functioning of the facial nerve, also known as Bell's palsy. In the traditional medicine, Somalis applied heat compress and/or lightly fire branded the area with hot iron rod. A slightly mad person, or in the English idiomatic expression as having "loose screw" is also said *"hebel shinbir buu qabaa"*, literally as "shimbir inside him or her". A similar affliction is also known in camels whereby they behave strangely.

For a helpless and a grief-stricken, they found solace in attempting to share their worries and hopes with birds in the hope that they may transmit they message. For those who 'attempted to heave their chests (hopes) up to reach their dreams, but lacking the legs (means)', just in line with the other Somali proverb (*Laabna la kacay, lugona ma hayo*), they found companionship in certain type of birds. A true example of such a situation was immortalized in the Zeilai' traditional dance song about two lovers – one living in the historical coastal town of Seylac (Zeila) at Gulf of Aden, and the other (female) in India. Their communication or reunion through the use of the sailing boats was interrupted by the harsh rough sea weather that brings movement of dhows to complete standstill for a certain part of the year. Therefore, they wished to be carried to one another by birds, same as we today travel in aeroplanes. However, realizing this mode of transport was impossible to materialize; they found consolation in imagining a metaphysical reunion made possible by the birds. Here is their dialogue:

> The guy: "Oo flying and swinging bird, would you take me on your wings? And would you put me by the side of my beloved darling?"

> The girl: "Came already the people of Saahil,[53] as those from Sur[54] as well; so what is wrong with the dhows from Mumbai; do they have mechanical problem? I glanced at the sea, and felt pain in my abdomen; so what is wrong with the dhows from Mumbai; do they have mechanical problem?"

> The guy: Me too, I would have loved to travel back (to you); but it is the rough season that kept me stranded in the way".

In the following example, a Somali poet,[55] attempts to share his worries with a hoopoe in a famous poem which the he gives us an insight into the extent of the deterioration and degradation of everyday life, following a long period of drought (1927-1930) and complicated by years of conflict. He wanted to put across to the hoopoe that he was not alone in the suffering brought by the drought, but the whole country was under great stress:

---
[53] Saahil = This is meant the Swahili coastal areas of East Africa.
[54] Sur = A port in Oman.
[55] Ismail Mire, one of the military leaders and confidente of Mohamed Abdullah Hassan (better known to the Europeans as the Mad Mullah of Somaliland).

"Guuguuleyahow haddaad gu'ga u ooyaysid;
Haddaad moodday keligaa inuu gubayo Jiilaalku;
Ama aad gabooyada ku maran gama' la diidayso;
Giddigood addoomaha waxa gaadhay seben weyne..."

O Hoopoe, when you shed your tears, [56]
Crying for the rains of spring
And avoiding the lightest wink of sleep -
 Just because your craw is empty.
Do you imagine that you, and you alone
Are scorched by this drought?
No, a great disaster has befallen
All God's servants, every one of them-
A drought is come that leaves nothing in its wake....

Our country teemed in the past with a wide diversity of birdlife. They included raptors that scavenged and cleaned the land of carcasses.

Now, where have all the birds gone?

The stark fact is that many species were exterminated by the widespread, indiscriminate and irresponsible use of pesticides and herbicides. The problem started in the early fifties' of the last century with the introduction of dipping tanks for pesticide treatment of livestock. Stainless steel tanks designed for this purpose that have been, at a later period, improvised into masonry cemented tank were installed or constructed into the ground. They used to be filled with water, added with a pesticide and then the dipping process of animals followed. Dieldrin, DDT, endrin, aldrin and lindane are organochlorine pesticides

---

[56] Jama Mohamed, *The Political Ecology of Colonial Somaliland*, Africa, 2004, Vol. 74 Issue 4, p534-566, 33p,

that were used as sheep dipping chemicals to treat sheep ectoparasites. However, the unfortunate thing was that the poisoned water was left in the facility, uncovered and undisposed of properly. Then, in country, where the precious fluid is so scarce and frequented by severe droughts, many birds and insects drank from such contaminated water. Their death can be immediate, but if they escape quick death, the accumulation of such poisons in their bodies renders the species barren since some of the persistent chemicals affect their reproduction system. The use of poison chemicals was also extended to kill carnivores that kill livestock, such hyena, fox, lion, leopard, etc. The most common technique was to put some poison in some parts of an animal killed by a carnivore, resulting in the death of many other animals, including raptors.

Farm pesticides are also causing death of birds and insects including honey bees. The chemicals accumulate in their bodies and cause neurological impairment and reproductive effects.

The birds are becoming rarer in the skies above us. Where are the birds that used to fly in huge flocks in beautiful and breath-taking formations, barely seeing one crushing with another? With their decline in numbers, the rarity in hearing the beautiful and musical cries of the different birds reminds us of the sad description on the plight of birds caused by pollution and toxic chemicals in Rachel Carson's (1962) book *"Silent Spring"* – a book that alerted the western world to the lethal effects of pesticides such as DDT and their effect on bird life. The present situation that can be observed in our "quiet" skies is a true reflection of the descriptions narrated by the said author. Their jubilant voices and calls that are renewed during the spring and autumn are becoming less frequent then before.

In the past, the appearance of raptors soaring high in the sky with their graceful circles scanning for prey from above, and then descending on a location, planted hope in the heart of the weary pastoralist searching for a strayed animal, or even a family members who lost his way in the bush. According to one adage *"raq iyo ruux"* that literally means "finding it either in carcass or in life", the sight of such descending raptors used to ignite the feeling and hope as well, of coming close to an important information on the fate of the missing person or animal, and to take a decision on the next move – whether to resign and abandon the search, or to continue the arduous journey.

It is a great loss for us to witness the extinction of these birds. The enjoyment we used to get from the seeing their bright colours, distinct sounds and calls, and showy displays will be gone forever. Also with their disappearance, their many ecological functions, such as helping in dispersal of seeds and pollens, loosening the endocarp of seeds to facilitate easy germination, etc. will suffer. Culturally, the rich indigenous knowledge born from hundreds of years of human-bird interaction will be lost.

Date Palm (Phoenix Dactylifera)

"The house that has no dates, its people is hungry." — a prophetic hadith, [*Tirmidhi*]

This article was written by Osman Mohamed Ali, from Action in Semi-arid Lands (ASAL), an NGO based in Boosaaso.[57] It was rendered into English by the author of this book

Introduction

Date palm is one of the first fruit tree species domesticated by man. It is believed that it was first planted in Iraq (Mesopotamia) as early as 3000 BCE. (Zohary and Hopf, 2000). The cultivation of date palm has played a pivotal role in the culture, history and socio-economy of the peoples in the Near East and North Africa. Around 1500 date palm varieties has been confirmed, 630 in Iraq alone.

The early expansion of Islam into North Africa, Europe and central Asia was instrumental in spreading date cultivation into areas where the climate was conducive for its planting. Dates cultivation has also spread in south western Asia and North America during the last three centuries. Equally, there has been a continual increase in its production. For example, till 2001, the overall global dates production has been increasing at a yearly rate of 5% (from 1.8 m tons in 1961 to 5.4 m tons in 2001) [58]

The plant is adapted to the hot, desert climate and saline waters. It thrives well in the availability of plenty of water, although its water requirement varies in accordance with

---

[57] Osman Mohamed Ali, email: cuthmaan@gmail.com
[58]. By André Botes and A. Zaid Date Production Support Programme Updated by Pascal Liu, FAO

the seasons. Soil type, temperature, humidity, and winds determine the amount of water it will absorb. Every part of the plant is known to have some use.

The quantity of date cultivated in the Somali lands is insignificant and does not go beyond local consumption. Nearly 98% of the existing tree stands are confined to Puntland, and all that is harvested is immediately consumed during the summer months of June to August.

In this essay, we will explore and highlight on how date palm was first introduced into the eastern part of Somalia, the attempts made to develop this sector and the challenges facing it, and its potentiality to contribute to the overall economy and income of farmers.

History of date plantation in Puntland

It is difficult to say the exact location where and whence date palm was first introduced into the region. However, a number of elders interviewed by the author, believe that the small settlement of Gacan, situated between Laas Qoray and Durduri, was the first place where date palms was reported to have been cultivated as early as the 17th century. The geographical proximity, the similarity in climate, and centuries of commercial, cultural, and physical interaction between the peoples of Arabia and the Somali coasts, made the adoption and dissemination of date palm cultivation on the African side of the Gulf of Aden possible. Oman, Iraq (with its famous port town of Basra, mainland Yemen, and Socotra Island) were the commercial hubs that the communities of Puntland interacted most with. The high energy content of the dates on one side, and the harsh climate of the country, may have also contributed to its introduction and acceptance. From there, date palms spread in other locations such as Sayn, Geesalay, Tayeega, Xogaad, Runya, Xaabo, Kalahaya, Qoralaho, Comaayo, Xeela and Baargaal in Alula and Iskushuban Districts, and Karin, Xamur, Dameer and Galgala in Boosaaso District. The most popular date palm varieties that have in the past reached the area are Nimcaan, Farad, Suxaari, Sahri, Faaquur, Barni, Suqadari, Miro-jad and Masiili. Date farms are positioned where water is easily procured in good quantities such as along the banks of running streams, or irrigated by shallow wells.

The shortcomings in the succession and division of the immovable assets from deceasers to heirs such as the common delay in inheritance distribution among heirs, is one of the main reasons that the trees end in neglect and subsequently become unfruitful or die. Therefore, it is not an uncommon sight to see unproductive farms with aged and/or dead trees everywhere.

A conservative estimation on the areas cultivated with date palm is 273 hectares with no less than 181,000 trees (SORSO, 1997), consisting of various ages ranging from aged and non-productive to young trees. Because of poor farming practices, the density of the trees per one hectare is 600 (four times more than the recommended number, 120-140 trees). Date palms trees should be planted in rows, 7- 8 m apart, with the same distance between two trees.

The **second phase** in the history of date cultivation in Puntland starts from 1984 when Somali's Ministry of Agriculture began an ambitious agricultural project that was thought to suit the history, climate, seasonality, soil and water in the region. Date palm, fodder production and farm animal improvements were the chosen areas of intervention. The project was financed by the European

Union countries (France in particular), and its main components were trainings on better agronomic practices including the selection of quality trees for offshoot propagation, enhancement of fodder production, in terms of quality and quantity, improvement in the irrigations systems, and introduction of the Somali-Arab Goat - a cross-breed originating from the Arabian Peninsula as believed locally. It has long hair and considered as a dairy Goat. It is mostly found in the southern coastal towns.

Boosaaso, Iskushuban and Alula districts received most of the project inputs. The date palm component, however, did not show much improvement as the cause of the challenges that have caused the old farms to be lie idle, due to succession and inheritance problem, remained unaddressed. Most of the farms had been crowded with old or dead trees. This also meant that no new farmers took up date palm farming as an occupation so as to plant first generation trees. One important development was that date palm offshoots from Bari Region was introduced to Halin (Xalin), an inland site, hundreds of kilometres away from the coast, and other sites as well for experimentation purpose. However, this initiative did not bear fruits due to the disruptions caused by the civil war.

During the **third phase** in the history of date palm cultivation in Puntland, this sector has seen a big revival in date palm production. The initial credit goes to Somali Relief Society (SORSO), a nongovernmental organization based in Boosaaso that exerted great efforts in raising community awareness and capacity building on date production. Most of the staff of the organization was former staff of the Ministry of Agriculture of Somalia.

SORSO established working relationships with international organizations and UN agencies operating in the region in procuring and bringing high quality tissue culture date palm from Al-Ain in the United Arab Emirates (UAE). The project started with a first consignment of 3000 trees and until 2008, the number of trees imported were 8,150, air-freighted and unloaded at Boosaaso airport. Between 2011 and 2015, another consignment of 2,100 tissue culture date plants, aging between 18-24 months, were imported by Action in Semi-arid Lands (ASAL), another local NGO based in Puntland. Again, this was followed by a bigger consignment of 40,000 young trees brought by the Red Cross Society. All this quantity, totalling 50,000 of imported trees, were planted in Bari, Nugaal and Sanaag regions. It is envisaged that, during the coming five years, an estimated 200,000 offshoots can be separated from those imported mother trees, saving money that could, otherwise, have been spent in procuring more plants from abroad.

The imported varieties differ from those that had been introduced in the past. The new varieties include Barxi, Majhuul, Khalas, Zamli, Shishi, Saqci, Umu-Dihaan and Suldaana, all procured from one nursery in the UAE[59]

In terms of production, an eight-to-fifteen year old tree can have yearly yields ranging between 60-100 kgs of fruits. Therefore, the 50,000 trees could potentially produce 5,000 tons per year. This is still far less than the regional demand for dates, bearing in mind that 3,293 tons of dates was imported through Boosaaso port during the period from June 1996 to February 1997.[60] In order to enhance

---

[59] Al-Wathba Marrionate.
[60] *WFP 1996, 1997*

production levels per tree, each tree should receive 5 to 20 kg of natural fertilizer per year

Because of the erratic rainfall patterns, shallow wells recharge levels and stream flows vary from season to season, and from year to year. This challenge is also compounded by the high cost of fuel for pumping water to irrigate date farms. Higher cost of production affects the profit margin of the farmers.

Ideally, date palm thrives in the *Guban* ecological zone, and if enough water to sustain it is procurable, it could be a fairly profitable and sustainable form of production, if it is adequately invested in. Therefore, this is a viable opportunity for local investors. However, in the light of the prevailing practices in resource use, characterized by heavy deforestation with its negative effects on watersheds, and the subsequent decline in the water table, cultivating and caring for trees with high water demand, does not seem an encouraging prospect, unless concerted efforts are taken to mitigate the on-going resource depletion.

A short description of other palm species found in the region

There are several palm species that are related to the date pam. These include Mayro *(Phoenix reclinata)* Cawsan/Qoome *(Hyphaene thebaica)*, Maddaah/Daabaan *(Livistona carienensis)*, and Aw-baar (Caw-baar), all of them possessing multiple socio-economic and ecological benefits. Mats, utensils, and robes are made from their fibre. The poles are used in buildings, while the fruits are eaten and the resin collected from some species are used for different purposes. Daabaan/Madaah is known for its height and scarcity. The few areas it still remains are Tasjiic, Geesa-qabad and Galgala in Boosaaso district, as well as in Dul and Maraje in Dhahar area – all located in the Golis Range. In Bosaaso, Elayu, Las Qoray, and other coastal towns, Daabaan was the cut in the past to serve the cross beams in buildings. The trunks were also hollowed for use as water pipes to carry rainwater out of buildings. It is believed that Daabaan poles were exported to the Gulf towns as construction material, along with Dhamas (*Conocarpus Lancifolius*), before the oil resources that revolutionized the Arabian landscape was struck. It is now estimated that not more than 100 Daabaan trees remain in Puntland. Therefore, there is an urgent need for an action for reviving it from near extinction. Cawsan and Mayro are two species mainly found in the lower coastal areas, along and within the seasonal water courses. Both species are widely used in the production household goods and materials. However, local skills are becoming scarcer, as many products of similar use are being replaced by imported cheaper materials. The fruits of Cawsan are known as "Xeego". These are white coloured fruits, internally intertwined with fibres. Eating its fruit is not easy, as one has to scratch his/her frontal teeth on the hard fruit. Hence, the famous Somali proverb "How best can Xeego fruit be scratched, and teeth can be free from damage?" {*Sidee loo xagtaa, ilkana u nabad galaan*}[61]. Mayro bears a fruit called "*Cawaag*", similar to that of the date palm. It is also found in the Nugal valley and Dhooddi seasonal watercourses.

---

[61] Here is another Somali proverb that conveys the same menaing: {*Si xeego loo ciyaaro, ilkona u nabadgalaan*} or "How best can a hockey be played, and teeth can be safe from damage".

28

Domestication of Henna in Somaliland
Botanical name: *Lawsonia inermis*

Geographical distribution

Henna is native to many parts of the Old World particularly in the Somali region, Sudan, Yemen, Egypt, Iran, India, Syria, Niger and others. In Yemen and India, the plant is also extensively cultivated. It is planted as hedges around houses, buildings and sometimes in fields.

Uses

The leaves, stems, flowers and seeds are used for cosmetic as well as medicinal purposes. In addition to its cosmetic uses as hair, nails and skin dye, it is also popular in the Islamic Prophetic Medicine (*Dibbu-Nabawi*), where its uses and benefits are well documented. It is helpful in relieving headache; its smell is said to be a nerve stimulant. It dries wounds and helps in the formation of healthy flesh (tissue granulation) as the wound heals. A gargle made from boiled henna leaves is beneficial for all ulcers of the tongue, cheeks and lips. When a tea prepared from its leaves is taken it can be helpful in stomatitis. If the paste of henna is applied on feet and soles, it is effective in burning feet syndrome. It has been employed both internally and externally in jaundice, leprosy, smallpox and various diseases of the skin. Dry leaves of henna if wrapped in clothes, serve as an insect repellant. The cut branches are resistant to pests and can be used by farmers to support tender plants such as tomatoes.

Economic benefits and marketing opportunities

From ancient times, henna has been utilized as a cosmetic dye for hair, skin and nails and it has acquired a particular significance in Islamic culture. There has been an increase in its usage as a hair and skin dye in Western Europe and North America in recent years. Prior to the widespread availability of synthetic dyes, henna was also employed as a dye for textiles and leather.

Henna bushes are found wild in the mountainous Golis Range area of Somaliland and many parts of Somalia as well. But despite its abundance, until a decade or so, most of the henna powder used in the Somali region originated from Yemen. For the first time, the domestication of the plant (from the wild) was carried out by the author of this book, who was also instrumental in establishing Asli Mills (a company under the umbrella of Candlelight organization) in 1998. Since then, tenths of thousands of plants were sown in farms at Adadley and Aw Barkhadle areas, and became a source of employment for tens of farmers, reducing import of henna into the country and contributing to saving the flight of hard currency out of the country.

Propagation of henna

Henna plants may be propagated both from seeds and cuttings.

　　1. *Growing Henna from seed*

Henna seeds require frequent watering for optimum germination (every other day when germinating in nursery bags, and on daily basis in seed beds). Nursery bags are the best for the propagation of henna plants. Fill the bags with a mixture of sand, dark soil and manure mixed in equal parts. The seedpods look like soft, round pouches that contain, depending upon the size of the seedpod, from 50-100 small, triangular, brown seeds. Put 5-10 seeds covered by soil into each bag. When germinating henna in seedbeds, seeds can be broadcast thinly onto the seedbeds and flooded with water regularly. The germination process is slow and it may take as long as 25 days, so one has to be patient to see the results.

2. *Growing Henna from cutting*

Another way of planting henna is by cutting portions of live sticks from branching parts of henna plants into lengths of 25 cm that are not more than 7 mm in diameter. The cut sticks can then be put into seedling bags or directly into the ground at a depth of not more than 8-10 cm. The survival rate of cuttings is only 25% compared to 80% when germinating plants from seeds. Seedling bags are best for planting henna cuttings. Water your cuttings every other day for healthy growth until they branch out (or for 4-6 weeks) and ready to transplant.

3. *Transplanting Wild Henna plants*

Small plants (less than 30 cm tall and 1 cm in diameter) uprooted from their habitat can easily be transplanted in another area. Make sure that cuttings and small uprooted plants are wrapped in a dampened bag (not plastic) in order to keep them moist until they are transplanted in a new location. Water your transplanted wild plants every other day for 2 weeks. The plants will shed their old leaves and once they bring forth new leaves the water regime could be reduced to once per week. Plants are mature enough to withstand longer period of watering intervals (3-4 weeks) when they start producing new branches and look well established.

4. *Establishing henna plantation*

When planting seedlings or saplings of henna into plantations, the plants must be spaced at a distance of 1meter in all directions. Seedlings/saplings are ready to be transplanted when they are 20-30 cm tall and .4 cm in diameter. Dig a hole 20-30 cm deep and 20 cm in diameter to plant your saplings/seedlings. Put layers of dry organic matter (leaves or grass), rotted manure and soil in the hole to help the trees grow well for a long time. Henna is a drought resistant plant so the recommended frequency of watering mature plants in rain-fed areas is between 25-30 days or longer. With frequent watering (once a week) and heavy treatment with manure, harvesting can be done 3-4 times per year. It is also important to note that as the root system of henna plants becomes well developed, and the older the plant, the less water it may require. Henna plants require little or no fertilizer, survives on rainwater, as well as little specialized labour.

Pests and Diseases

In the drier rain-fed areas, young henna seedlings and cuttings are vulnerable to termites that are attracted, particularly during the dry season, by the moist micro environment around the newly planted cutting. The termites devour the bark of the stressed plant causing them to die. The problem of termites can be avoided by frequent disturbance of the soil around the cutting (*baaqbaaq*). Otherwise, henna plants have great resistance to other pests. In case of pest infestation, chemical spraying is not recommended - since henna will eventually be applied onto the skin. Instead, bio-pesticides are recommended. Examples of appropriate natural pesticides are tobacco leaves, neem leaves, *Salamac* (*Cadia purpurea*) leaves and myrrh (*Commiphora myrrha*). These can be ground up and soaked in water overnight before spraying on affected plants. Ashes can also be applied to deter pests.

Harvesting

The best time to harvest henna is when there is the least chance of the leaves being dampened by rain. Early July crop is the best when the dye levels peak due to the comparatively higher heat and the January-February crop is the weakest.

The plant is coppiced and cut about 5cm above ground level. Usually one shoot is left uncut so as to facilitate the plant to recover from the damage. The cut leafy twigs must be dried in a sheltered place protected from direct sunlight and wind for 4 – 5 days. When the leaves are dry they can be shaken from the branches by striking it with a stick. Next pack the dry leaves in bags ready for processing but make sure that the leaves are completely dry as semi-dried leaves could be prone to mold formation, color change, and eventually produce lower quality dye. The advantage of coppicing over direct leaf collection is that one will get better quality henna. Henna from pruned plants has greener leaves and the end product has a deeper red color when diluted with water. Plants pruned correctly will regenerate from the cut area bringing forth multiple shoots.

Drying and Storage

Leaves must be dried in the shade, protected from the direct heat of the sun and wind to keep them green. Customers prefer henna to be very green. The shelf life of properly dried henna is three years. Then the leaves are finely ground into powder. Consumers do not like coarsely ground henna; therefore, the fineness of grinding is very important. Make sure that the leaves are free from stones, sticks, manure, henna seeds or any other foreign bodies that might find the way into the raw material. Sieve the final product to ensure the fineness of the powder. This will make your product more attractive to your customers even if you lose a small amount as 'waste' in the form of leaf fiber. The organic residue need not be wasted. It can be used as a fertilizer and for pest control. Spreading the residue around the root area and mixing it with the soil serves as pest repellent. Through experience, it was found to be useful in discouraging termites from damaging the lower stem are. The twigs can be used as cattle feed, and as shade material for *Berkads* (in-ground cemented water tanks) in order to reduce evaporation.

Miscellaneous Environmental Reflections

Displaced bees

One day, during a trip to the countryside, I came across an area whereby charcoal production was going in earnest. Then my eyes caught the numerous mini pyramid-like mounds, containing below their surfaces the 'limps' of butchered trees, covered with combustible herbage, and finally a layer of flattened iron sheets spread over the wooden stacks. On each of those structures, another layer of soil was put over them to make the process leak-proof, and limited the amount of oxygen seeping into the material, with the aim of kilning the ignited wood. A small hole at the top of the heaps served as an escape vent of the smoke generated by the burning wood. The vent is closed when the carbonation process is completed. Looking around, I saw how the surrounding landscape was turned into a waste. There was not a single misused tree in sight. Tree stumps with fresh axe marks and thorny branches severed from tree trunks, their leaves turned yellow and dead, was not a pretty sight to look. Bird nests with broken eggshells were loosely hung from the branches lying on the ground.

My mind raced back many decades to my childhood days. This area was a '*kob*', thickly covered with luxuriant vegetation. The soil was rich, the grass was matted, and the plentiful shade kept the soil moist for an extended period of time after the rains. The abundant vegetation was laced with *armo* creepers, making the bush an ideal hiding area for carnivores. Fear and caution used to creep into my head and heart whenever I approach the bush, especially when I was in search of my family's two burden camels, before the darkness creeping from the east could chase the twilight to follow the footsteps of its exhausted mother – the Sun, thereafter, shrouding everything under the roof of the sky with total darkness.

Further down the road, I reached Magaalo Yar, whereby I stopped for a cup of tea, in order to regain some energy and, at the same time, dilute the bitterness inside me and wash the dryness in my throat, left behind by the scenes of the denuded landscape. Hungry bees feasted themselves gluttonously on my tea. One of them made a painful sting to my upper lip. Before I finished my tea, I noticed a small crowd of people around me. The villagers thought I could be of some assistance in solving a problem that they were facing. They were under the constant attack of angry swarms of bees. They told me that was the first time in the memorable history of their settlement to have experienced such large swarms of ferocious insects. They knew I was representing an organization that carried out beekeeping projects.

I told them they were paying the price of ruining the homes of myriads of insect species - including bee, by destroying their habitats through charcoal production, and reducing them to state of displacement, destitution and death. I added, "even if we catch the swarms and put them to hives, they will die of starvation, as the source of pollen and nectar is being drastically diminished day after day".

Nutrient cycle

Plants need nutrients from the soil to grow. Soil nutrients mostly come from the breakdown of mineral-bearing rocks and from organic matter resulting from the decomposition of plants and animals. The nutrient cycle is a concept that describes how nutrients move from the physical environment into living organisms and subsequently is recycled back to the physical environment.

In my childhood, I had a preference for evergreen trees over the leaf shedding ones. I believed that they were better than the deciduous trees that can be seen bare during the two dry seasons. Later, however, I came to know through my botany classes, that they were as good as the evergreens. Now, there is a fig tree (*Ficus vasta*) in my house compound (probably the first and the oldest of its kind to be introduced into the city of Hargeysa). Each year, it sheds an average 120 kg of leaves. I then collect and transport them to my farm located at a distance 90 km away, with the aim of enriching its soil. Reaping such benefit from my generous tree made me realize that, contrarily, the evergreens are 'miser'.

The only drawback of my fig tree is that it causes a daily broom work, but still worth the effort. The litter helps maintain the nutrient cycle and hence, its productivity.

Adaptation

Plants do not have brains but are full of chemical reactions such as photosynthesis. Therefore different stimuli (like sunlight, temperature, fire, getting browsed by livestock, such as camels) trigger chemical responses in the plants.

Those responses can cause the plant to do all sorts of things, like grow taller, drop leaves, equip themselves with different types of armoury, such as thorn, pungent smell, toxic chemicals, to serve as a protection against browsers, or by making themselves unpalatable.

There is an acacia species locally known as "*Jaleefan*" (*Acacia hamulosa*) that is adapted to harsh desert environment of the coastal (*Guban*) ecological zone. It is very much similar to the *Bilcil* (*Acacia mellifera*), mainly found in the cooler *Hawd* plateau. The leaves of this plant are armoured with minute hooked thorns. The peculiarity of the Jaleefan is that its tiny leaves have miniature hooky thorns at their backs that can brick the tongues and the other soft membranes of the mouth of browsers.

The Eastern Horn of Africa has been undergoing a state of progressive aridity for the past several millennia, and equally the local flora has been adapting to these harsh environments so as to thrive and survive. Equally, herbivore adaptation to plant defence continued, often resulting in reciprocal evolutionary change. .This could be compared to the modern-day "arms race".

Unsightly graffiti on Sheikh Mountain Road

The picturesque and aesthetic beauty of the Sheikh Mountain Pass is being soiled by graffiti on rock faces – most of them in the form of business advertisement. For a nature lover like me who, from time to time, opts to disappear into the wilderness and take a break from the hustle and bustle of the city life, the ugliness and untowardly presence of these paintings in a location they do not belong is irritating. Moreover, plastic shopping bags and water and soft drink bottles litter the road side.

The place must have been part of the land that Queen Hatshepsut, in her lofty expression of homesickness, sang the following:

> "It is the sacred region of God's Land; it is my place of distraction; I have made it for myself in order to cleanse my spirit, along with my mother, Hathor...the lady of Punt."

After a period of two millennia, and with all the ravages of climate changes and decades of accelerated land degradation, the place still carries a portion of its beauty and mystique. Isn't it unfair that we leave such negative foot print on it, and definitely cause the outcry of the generations yet to come about our irresponsible actions?

The Lone Tree

There stands in the middle of the great Bancadde plain, in Sanaag region, a lone Higlo *(Cadaba heterotricha)* tree. The first thought that comes to mind is how did it survive, thrive, and escaped from the all-time axe wielding pastoralists in the first place? It carries all the hallmarks of a survivor. In country where droughts are a recurrent feature, the seed must have sprouted in year of good rain. During its tender years, it must have been lucky to be protected from being uprooted by goats amid the shelter of shrubs. Then it grew to some height whereby it became exposed to the nipping of the browsers. It fought back through the activation of its alarm system to produce chemicals to discourage herbivores. Years later, it grew to a height whereby the pastoralists might entertain cutting it for use in animal enclosure *(Xero)*. However, the need for its shade against the heat of the summer days counterweighed that risk. In another couple of years, the community bestowed a name on it that spread far and wide. A tree like this serves as a land mark, a shade for people, livestock and other animals, a venue for meetings and for women to organize their *Sitaad (Abbay abbaay)* sessions, a place children play their traditional games such as *Leelo goobaley* and *Shabadaan*, and a station for perching birds and for their nesting. Cutting such a tree was reacted with abhorrence. In reconstructing mental images in the past, when Man lived more harmoniously with nature than these days, the following picture can be illustrated: A herdsman and his flock of sheep rests under the shade of this tree, while a giant lizard (*Maso-cagaley*) escaping the relatively elevated ground temperature seeks rest in the coolness of its branches where it remains for most part of the Summer season. The descending of the reptile to the ground always carried an important message as it ushers the approaching of the rainy season.

Avian Tragedy
Day after day, more buildings with tinted glasses (mainly of sky blue colour) and aluminium wall panels are crowding our cities. They make our cities beautiful, but on the negative side, these buildings are becoming the killing grounds of many bird species as they mistake the blue 'barrier' for an open sky. A friend of mine alerted me to this 'unintentional massacre' as he had witnessed the headlong crashing of 27 birds in one building in a period of one month. Very sad!

God is the splitter of the seed-grain
"It is God who causes the seed-grain and the date palm to split and sprout. He causes the living to issue from the dead, and He is the One to cause the dead to issue from the living. That is God: Then how are ye deluded away from the truth?" Holy Qur'aan: 5:95)
What a joy it makes for someone among us to act as the medium that causes a dormant fruit seed, left on our dinner table, to sprout, grow, provide shade, bear fruits, absorb carbon dioxide from the atmosphere, and exhale the life-giving oxygen, and finally expect God's reward in the Hereafter!

Warthog

The extinction of the lion, the main predator of the warthog (*Phacochoerus aethiopicus*), caused the increase in the population of the later. Equally, Baboons (*Papio hamadryas*) have also been multiplying in number, as the leapoard, its main predator, has been decimated through pouching. Had both animals not been sanctioned as unlawful (*Haram*) by Islam, their number could have been drastically reduced. On the other hand, the ecological benefits of the warthog cannot be underestimated. Their role in breaking the soil through trampling (hoof effect), and defecating, like other ungulates (hoofed), contributes to water effectiveness, and soil enrichment.

Termites

While termites are commonly viewed as pests, they have an important ecological function. They are actually great decomposers. Termites break down tough plant fibres and wood, recycle dead and decaying trees into new soil, thus enriching it and increasing water infiltration into the ground.

Earth's resources

"The earth has enough for everyone's need but not enough for everyone's greed."

Mahatma Gandhi said this at a time India's as well as the world's population was not growing as alarming as today's. Now because of the equally alarming rate of resource depletion and population explosion, we can now boldly say that the earth has measured resources for a limited number of thrifty and environmentally conscious people.

From the pages of Drake Brochmann's "British Somaliland" (1910).

".... but a reward awaits the keen botanist after a few showers; the whole scene is changed as though some magic wand had been passed over it, the trees in a few days are more or less covered with leaves, and in another week or fortnight they are in flowers, and the air is laden with a scent of the acacias; while the smaller plants and bushes are vying with each other to be the first to brighten the landscape with their colours, the quaint horn bills, the brilliant starlings and babblers, and the larks, enliven the bush and plains respectively with their chattering and their songs; while the moonlit nights on the banks of the larger rivers where there are tall trees are cheered by the sweet notes of the Somali nightingale. If this is a desert it is a very pleasant one."

Now a little over 100 years, since this writer made this note, the country is now almost turned into a desolate waste. This is really a sad contrast with the past. The quality of life of the people has also taken a nosedive. For the sake of our existence, let us join hands to keep our environment, guard it, and care for it, for it keeps us, guards us, and cares us. Destroy it and we are destroyed.

Good intentions can sometimes lead to environmental disasters

It is a positive trend that different communities vie for self-help projects to develop their areas. However, the degree of success of such projects do not depend alone on own resource mobilization, community participation, the extent they touch and impact their lives alone, but also on good planning. Poor planning can kill good initiatives.

We will take the following story as an example: There was a community-led road construction project that the participating community mobilized some resources from their members. However, rather than completing the road in stages in line with the resources in hand at a given period, and then inching the progress to the next stage, all that money in hand was used in bulldozing the earth in a straight-line for an estimated length of sixty kilometres towards the destination point. The expectation of the planners was that the in-flow of resources will continue, as the 'progress' made could motivate more people to contribute money towards the completion of the project; but for some reason or another, the inflow of cash has faltered. Now imagine 60 km straight-line of earth cut to a minimum depth of 10 cm below the surface of a prime grazing land. The cutting and tearing of the bulldozed earth was like a camel slit at the throat, with its stone dead eyes facing the heavens. Millions of shrubs, trees and grasses have been cleared during the operation. Vehicle drivers immediately realized the cleared, unpaved path was a shortcut route, its tyres causing further breakage of the soil. Then rains came, and in couple of years, the dirt road turned into a seasonal watercourse. Inch by inch downstream, the flow of water accelerated carrying with them the fractured soil. The initiative that was intended to be an arterial route to connect villages, ease access and travel time, has all of a sudden turned into a mammoth snake draining the life-giving water and fertile soil from the nearby grazing land. One major drawback that can be added to the poor planning of the project is the absence of consultation with environmental engineers to investigate the probable impact of the project and then recommend ways to mitigate their effects.

Verily, good intentions can sometimes lead to environmental disasters.

Trees can also repay good!

Few years back, while driving in the rural areas, I came across a man putting fire at the trunk of a big *Kidi* tree (*Balanities aegyptica*), to make it fall and later cut it into logs to kiln it with the aim of producing charcoal out it. I stopped and immediately rushed to pour some water I carried with me onto the slow burning trunk and eventually succeeded in putting off the fire. This was followed by a heated argument starting like this: "Is it your tree that I am burning?" to which I replied; "It is our tree that you are firing". Finally, I succeeded in convincing him to leave the tree alone. Other than its ecological benefits, this tree is useful for making high quality poles for the Somali collapsible hit *(Aqal)*, axe handles *(daabka gudinta)*, knife handles, and the all-important *Hangool* (the pastoralists' forked stick). It is resin is used as cement to leak-proof water and milk containers. Its gum is also said to strengthen the teeth and gums.

Another couple of years later, while driving on the same road, one of the tyres of my car ran out of air. I stopped to examine it, and then repair it. Then I straightened my back while, at the same time, wiping profuse perspiration from my forehead. Finally, I turned my eyes around me, and behold! It is the very tree that I saved from the charcoal producer. I walked towards the shade of the tree in a state of excitement and happiness. I stood under its shade and it was as refreshing as what seemed to be a breath from heaven passing over my forehead. I started hugging the tree, and has it had movable arms, probably I could have felt them wound around my back.

Souls of Believers in Green Birds

The Messenger of Allah (SAW) said, "The souls of the Believers are inside GREEN birds in the trees of Paradise until Allah returns them to their bodies on the Day of Resurrection." [At-Tabraanee]

This is strong message reminding us to the importance of protecting and conserving the environment.

Trees are one of God's first creations. Modern science also proves that primitive plants converted the poisonous atmosphere of the early earth into one rich in oxygen, able to support animal and human life. In Islam, Paradise is akin to a garden <Jannat u adn> (Gardens of Eternity), and in Christianity it is the <The Garden of Eden>. Prophet Muhammad stressed on the importance of tree planting in his famous Hadith **"If the Hour (doomsday) is imminent, and anyone of you has a palm shoot (to plant) in his/her hand, and is able to plant it before the Hour strikes, he/she should do so, because he/she will get recompense (from Allah) for that action"**.[62]

The 'benefits' of trees to people do not end with their death. The custom of planting trees has also spiritual significance. In another Hadith, Prophet Muhammad, while passing a graveyard, the narrative mentions that Allah revealed to him the torment of the occupants of two graves, one because of backbiting and other because of the impurities while urinating. Then the prophet took palm shoots and one on each grave, also mentioning that such

---

[62] As-Silsilah as-Saheehah #9., by Shaikh Al-Albaani

act will serve to reduce the pain the occupants suffer in the *Barzakh* (Purgatory).

A scene in Gallaaddi (Galadi), dating over a century
"......At last we came on the oasis called Galadi, a very remarkable place, set like a jewel in a rim of iron. We could hardly believe our eyes. It was such a faceted gem. No more dingy brown landscape, but a peaceful sylvan scene of great trees, real turf, and a wealth of green vegetation. This patch of emerald extended for a mile or more and seemed like a little Heaven." *(An excerpt from the Book, The Two Dianas of Somaliland - 1908).*
It is said that a picture can tell 1000 words, but I believe this piece can tell more than 1000 pictures. I am keen to know how far the above description could be different from the current situation of Galadi.

Cimilo: Somali word for "climate"
The Somali word for 'climate' is 'cimilo'. Do you know how this term came into existence?
A friend of mine living in the UK who attended one of the presentations of the late great Somali poet "Gaarriye" told me that the poet informed them that the word originated simply from the misspelling of the English word 'climate'. Gaarriye was a member of the Somali Language Committee that was established in 1972 culminating in the introduction of a national orthography for Somali in Latin script. He said he told the audience that a lady who was then a secretary of the Committee misspelled the word in the form of 'cimilo' where after the committee decided to use the word for 'climate'.
Thanks to our late prolific and sagacious poet and also to my friend (*Abdirahman Abtidoon*) for sharing this with me.

Noise pollution

Sound is essential to our daily lives, but noise is not. Noise can be defined as unwanted sound. It is a source of irritation and stress for many people and can even damage our hearing if it is loud enough. Noise pollution is a type of energy pollution whereby distracting, irritating, or damaging sounds are freely audible. As with other forms of energy pollution (such as heat and light pollution), noise pollution contaminants are not physical particles, but rather waves that interfere with naturally-occurring waves of a similar type in the same environment.

A popular feature in the streets of Hargeysa is the unnecessary hooting of horns by drivers and high bitch noise emanating from the whistles of the traffic police. The high pitching hooting of cars, *Qat* transporters in particular, is very disturbing and sometimes cause accidents. Many people find this disturbing, distracting and irritating. It is a form of environmental pollution.

You must practice what you preach

One day, during my employment with Candlelight, a prominent environmental organization operating in Somaliland, a visitor from another organization that funded a project being implemented at that time our organization arrived to pay a monitoring visit. In my capacity as the Executive Director, I decided to accompany her to the field visit. Two other staff, one of them a junior member, also joined us. On our way to the field, we found it a good opportunity to brief her and elaborate on the multi-tasks of the organization in the environment sector as well as the pro-activeness and stewardship of its staff.

She became so much impressed with the narrative. Then all of a sudden, the junior staff member, who was sitting beside the foreign visitor, hurled a soft drink bottle, after emptying it, out of the car window. The reaction of the visitor was immediate and straightforward. "How would you reconcile this action with your environmental stewardship?" She said disapprovingly but also jokingly.

We all looked at one another in astonishment. Matter of fact, there was no words to justify such environmentally unfriendly action.

Remember, one must practice what he/she preaches.

PLATES

Plate 1

Plate 2

Plate 3

Plate 4

Plate 5

Plate 6

Plate 7

Plate 8

Plate 9

Plate 10

Plate 11

Plate 12

Plate 13

Plate 14

Plate 15

Plate 16

Plate 17

Plate 18

Plate 19

Plate 20

Plate 21

Plate 22

Explanatory Descriptions of the Plates

<u>Plate 1</u>: *Warab-ka-roon /Habartacay (Adenium somalense var.):* A rare species found in the Golis mountain areas. Fibre is extracted from its bark for making containers and robes. Location Karin Geeltaba, north of Gudmo-biyo-cas in Sanaag region *(Photo by: Ahmed I. Awale)*

<u>Plate 2</u>: *Madaah / Daabaan (Livistona carienensis)*, Gudmo-biyo-cas, Sanaag Region. *(Photo by: Ahmed I. Awale)*

<u>Plate 3</u>: *Adaahi/Habartacay (Adenium somalense)*, Marso area, west of Sheekh, Sahil Region. *((Photo by: Ahmed I. Awale)*

<u>Plate 4</u>: *Berde (Ficus vasta)*, Wadaamago, Togdheer Region. *(Photo by: Ahmed I. Awale)*

<u>Plate 5</u>: Dacardhaadheer *(Aloe eminens)*, Tabca gorge, Daalo, Sanaag Region. *(Photo by: Ahmed I. Awale)*

<u>Plate 6</u>: *Maacaleesh (Delonix regia):* Indiscriminate cutting of Flame tree *(Delonix regia)* in Hargeisa *(Photo: Ahmed I Awale)*

<u>Plate 7</u>: *Maacaleesh* (Delonix regia): Flame (Flamboyant) tree in one of the streets of Hargeysa, cut indiscriminately. *(Photo: Ahmed I Awale)*

<u>Plate 8</u>: Lichens (*Usnea articulata*) hanging from *Juniperus procera* tree in Daalo, Sanaag Region. This is an indicator of presence of mist in reasonable quantities. *(Photo: Ahmed I. Awale)*

Plate 8: Dragon Blood Tree (*Draceana schizantha*) hanging from Daalo escarpment, Sanaag Region *(Photo: Ahmed I. Awale)*

<u>Plate 9:</u> The author standing by a Dragon's Blood tree hanging at the edge of Gacan Libaax escarpment. In the background, the gradient continues to fall into Guban lowland and coastal maritime plains.

<u>Plate 10:</u> An excellent stand of perennial grasses surrounding a *Cupressus spp*. This area was almost a bare

ground prior to the conservation work carried out by Candlelight starting from year 2000. (*Photo: Ahmed I Awale*)

Plate 11: *Waxaro-waalis (Ipomea cicatricosa)*. Beautiful desert flower *(Photo: Ahmed I Awale)*

Plate 12: Serious watercourse bank erosion, Adadley, Maroodi Jeex Region. (*Photo: Harir Ibrahim*)

Plate 13: Remnant Somali horses at Bancadde plain, Sanaag Region (*Photo: Ahmed I Awale*)

Plate 14: Reaching out for the precious fluid from a surface water catchment (*Balleh*) in Togdheer region, South of Burao. *(Photo: Abdirizak Bashir)*

Plate 15: Charcoal ready for shipment from Kismayu Port (2003) (*Photo courtesy: KISIMA Org.*)

Plate 16: A man-made vent in Tabca Pass, Daallo Mountain. The wonderful Taba' gorge is in the background. *(Photo: Saeed Nuh)*.

Plate 17: Winter sun setting over Nasa Hablood, Hargeysa, Somaliland (*Photo: Ahmed I Awale*)

Plate 18: Dust storm over Nasa Hablood created by an oncoming rain. The greater the deforestation, the more intense storm episodes that are experienced. (*Photo by: Ahmed I. Awale*)

Plate 19: Somali Wild Ass, (St. Louis zoo, 2005). *Photo by: Robert Lawton (Source: Creative Commons)*

Plate 20: Swayne's Hartebeest or Somali Hartebeest (*Alcelaphus buselaphus swaynei*). It has been wiped out during the middle of the 20th century. The only thing that could remind us is a seasonal watercourse near Adadley town that carries the same name.

Plate 21: Date palm tree, Dhud, Bosaaso, Puntland (*Photo: Cismaan Maxamed Cali*).

Plate 22: Mats made from Caw (*Hyphaene thebaica*) ready for sale, Dalweyn, Qardho, Puntland. *(Photo: Cismaan Maxamed Cali).*

List of Common Plants
Botanical Name | Vernacular Name

Acacia albida	Garbi
Acacia hamulosa	Jaleefan
Acacia horrida	Sarmaan
Acacia bussei	Galool
A. edgeworthii	Jeerin
Acacia etbaica	Yubbe/ Sogsog
Acacia millifera aff.	Bilcil
Acacia misera Vatke	Qansax
Acacia nilotica (Linn.) Del	Maraa
Acacia orfota	Gumar
Acacia Arabica	Cadaad
Acacia tortilis	Qudhac
Acokanthera Shimperi	Waabay
Adenia veneata	Dhallaanlaaye
Adenium Somalense	Warab-ka roone (Habar-tacay)
Aeluropus lagopoides	Afruug
Agave sisalana	Xig dhaadheer
Aloe eminens	Dacar dhaadheer
Aloe socotrina	Dacar dhegweyn
Amaranthus blitum Linn.	Cayo guri
Andropogon kelleri	Duur
Andropogen sp.?	Gogane
Asparagus recemosus	Argeeg (ergeeg)
Avicennia marina	Takhay
Azadirachta indica	Niim (Mirimiri Hindi)
Azima tetracantha	Qodaxantoole

Balanites orbicularis	Quud
Balanities aegyptiaca	Kidi
Balanites spp.	Kulan
Berchemia discolor	Dheen
Blepharis edulis	Caraancar
Blepharisspermum fruticosum	Gahaydh
Boscia coriacea	Qadow
Boscia minimifolia	Meygaag
Bowellia carterii Birdw[63].	Moxor lab
Boswelia frereana	Yagcar, Moxor cad[64]
Boswellia sp	Muqle
Brachiaria leersioides	Cagaar
B. ovalis	Baldhoole
Buxus hildebrandtii	Dhosoq
Cadaba farinose	Dhiitacab
C. heterotricha	Higlo
C. purpurea	Salamac
Caesalpinia sp.?	Jirma
Calotropis procera	Booc
Caralluma socotrana	Gowracato
Cassia obovata	Jaleelo
Casuarina equisetifola	Showri
Celtis kraussiana	Dhebi boodaar
Cenchrus ciliaris	Guddoomaad
Cephalocroton cordofanus	Jimbac

---

[63] The resin of the *Moxor* is "*Beeyo*". Arabs call it "*Luban Dakar*" (lit. male frankincense). Other than using it as incense, it is diluted with water and drank by persons suffering from venereal infections. False amber was also made from it, worn by women. The beads were threaded together into necklaces and rosaries. Men used to have a single piece rimmed with leather, and then worn on the neck.

[64] The gum of Boswelia frereana is "*Mayddi*". It has good gum. Its resin is also good to leak-proof water and milk containers.

Ceratostigma speciosum	Arabjab
Ceropegia sp.	Marooro
Ceropegia sp.	Doombir
Chrysopoden aucheri	Dareemo
Cissus sp.	Carmo
Citrullus vulgaris Schrad.	Unuun
Cocculus pendulus	Xayaab
Combretum sp.	Cobol
Commelina sp.	Baar
Commicarpus spp.	Geed irmaan
Commiphora candidula	Raxan reeb
Commiphora myrrha	Dhiddin[65]
Commiphora crenulata	Gowlallo
Commiphora drake-brockmannii	Dhunkaal
Commiphora erythraea	Xagar cad
Commiphora hildebrandtii	Xagar madow
Commiphora hodai	Xoday
Commiphora opabalsamum	Goron madow
Commiphora guidotti	Xabag hadi[66]
Commiphora opobalsamum	Dhasayno
Commiphora tubuk	Tobbog
Conocarpus lancifolius	Dhamas
Cordeauxia edulis	Gud (geedka Yicibta)
Cordia ovalis/ Corida somalensis	Madheedh
Crotalaria comanestiana	Gabbaldaye

---

[65] In the past, small amount of "*Malmal*" was diluted with water and the nursed with the new-born babies, also later during teething. In the first place, it was believed that it was good against the evil eye, and in the later, as painkiller.

[66] Small amount of *Xabag Hadi* was diluted with water for urinary tract infections. It was also administered on milch camels as a drink to improve its health.

Cymbopogon floccosus	Carrabjeeb
Cymbopogon schoenantus	Caws dameer
Cynodon dactylon	Doomaar meadow
Dactyloctenium robecchii	Gubaangub
Dactyloctenium scindicum	Saddexo
Dalbergia commiphorodes	Dhuyuc
Danthoniopsis barbata	Caws tumbulle (timoole)
Delonix elata	Lebi
Doberaa Glabra	Garas
Dodonaea viscosa	Xayramad
Dracaena schizantha	Mooli
Dasysphaera robecchii	Maraboob
Edithcolea sordida	Xamakow
Ehretia orbicularis	Himir
Ehrharta abyssinica	Gowsomadoobeeye
Eleusine floccifolia	Gargoor
Eragrostris papposa	Xarfo
Etada abyssinica	Faradheere
Euclea kellau	Maayeer
eucalyptus camaldulensis	Baxrasaaf
Euphorbia grandis	Xasaadin
Euphorbia spinescens	Derinder
Euphorbia gossypina	Cingir
Euphorbia robecchii	Dharkayn
Euphorbia somalensis	Falanfalxo
Euphorbia spp.	Dhalanyadhuux
Euphorbia breviarticulata	Dibaw
Ficus glumosa/ F. Vasata	Berde
Ficus salicifolia Vahl	Dhicir
Ficus sycomorus L.	Daray
Glossonema boveanum Decne.	Sobkax
Grewia bicolor Juss.	Dhebi
Grewia erythraea	Unidentified sp.
Grewia erythraea	Midhcaanyo

Grewia tenax	Dhafaruur
Grewia villosa Willd.	Gommoshaa
Grewia sp.?	Hohob
Gymnosporia arbutifolia	Dhegmud
Gynandropsis gynandra	Cayo guri
Gyrocarpus asiaticus	Yucub
Heteropogon contortus	Cawsguduud
Hildebrandtia somalensis	Daanyo
Hyphaene carinensis Chiov./Livistona carienensis	Maddaah
Hyphaene reptans Becc.	Caw
Hyphaene thebaica	Qoome, cawsan
Hypoestes forskaolii	Geedo waraabe (Faaraxood)
Indigofera sparteola	Jillab
Iphiona rotundifolia	Gagabood
Ipomoea donaldsoni	Bulumbul
Ipomoea nephrosepala	Jadeer
Ipomoea sp.	Waxarawaalis
Jatropha sp.	Jilbadhiig
Juniperus procera	Dayib
Justicia sp.	Sarin
Justicia minutifolia	Buuxiso
Kelleronia spp.	Gorayo kaga xaaris
Lachnopylis oppositifolia	Biyeys
Lactuca spp.	Burdad
Lantana petitiana	Geed Xamar
Lasiurus hirsutus	Darif
Latipes senegalensis	Jabi-oke
Leptadenia spartium	Moroh
Limonium axilllare	Caws biyood

Lowsonia inermis	Cillaan
Lycium europaeum L.	Surad
Maerua angolensis	Laamalooye
M. crassifolia	Jiic madow
M. sessiliflora	Jiic
M. sphaerogyna	Qadow
Ochradenus baccatus	Mirrow
Ocimum americanum	Dhikri
Olea somalensis Baker	Weger
Osyris abyssinica Hochst.	Casaaso
Panicum turgidum	Dungaari
Parkinsonia aculeate	Sabsabaan
paspalidum desertorum	Garagaro
Pavetta venenata	Ruqumbaa
Pavonia Arabica	Midho geeljire
Pavonia sp.	Balanbaal
Phoenix dactylifera	Timir
Phoenix reclinata	Maydho
Tamarindus somalensis Mattei	Xamar (Jaadla xerodhaladka ah)
Pistacia lentiscus L.	Xiis
Pistacia falcata Becc.	Siisay
Prosopis juliflora	Garanwaa
Psiadia Arabica	Weila subke
Raphanocarpus stefaninni	Gasangus
Reseda oligoeroides	Dhebi-yar
Rhus somalensis	Ilka caddeeye
Salsola bottae	Gowsomadoobeeye
Salsola foetida	Gulaan
Salvadora persica	Cadday/Rumay
Salvia somalensis Vatke	Surad madow
Sansevieria guineensis Wild.	Dacar dhegweyn
Sansevieria ehrenbergii	Xaskul
Sansevieria abyssinica	Xig

Agave sisalana	Xig dhaadheer
Sarcostemma viminale	Xangeeyo
Senecio longiflorus	Godor
Senra incana Cav.	Balambaal
Sericocomopsis pallida	Wancad
Sesamothamnus busseanus	Salelmac
Schinus molle	Mirimiri
Sideroxylon gillettii	Shooy
Solanum carense dunal.	Kariir
Solanum somalensis	Dameero rogad
Solanum sp.	Mooh
Solanum sp.	Micigoodle
Sporobolus spp.	Ramas
Sporobolus ruspolianus Chiov.	Sifaar
Sporobolus spacatus (Vahl.) Kunth.	Ris
Sporobolus variegatus	Dixi
Suaeda fruticosa	Xudhuun
Suaeda sp.?	Daren dameeraad
Suaeda vermiculata	Dinaas, Daran cad, daran
Tamarix aphylla	Dhuur
Terminalia polycarpa	Hareeri
Tetrapogen spathaceus	Aya mukarre
Themeda triandra	Dabashabeel
Tragus racemosus	Nagaadh
Trematosperma cordatum Urb	Mawo
Tribulus terrestris	Gocondho
Tricholaena teneriffae	Ramad gudhiye
Tripogon subtilissimum Chiov	Maxaansugi
Triumfetta sp.	Salweyne
Caralluma somalensis	Ubaatays

Usnea articulata Hoffm.	Jiibaan
Vernonia cinerascens	Hiil
Withania somnifera	Guryafan
Zizyphus hamur Engl.	Xamudh
Zizyphus mauritiana Lam.	Gob
Zygophyllum hildebrandtii	Aftaxolle
Zygophyllum simplex	Kabaqoys

For more comprehensive lists of Somali plants, please refer to the following documents:

1. *A Provisional Checklist of British and Italian Somaliland Trees, Shrubs and Herbs. Glover*, P. E., – 1947. Crown Agents, for the Government of *Somaliland*
2. The Flora of Somalia, 4 volumes. Kew Royal Botanic Gardens (1993), Mats Thulin
3. *The Flora of Somalia: Somali Plant Names and Dictionary (Qaamuuska Magacyada Dhirta Soomaaliyeed)*, Ahmed M. I. Barkhadle; Florence, May 1990
4. *Herbarium Collections at South Eastern Rangeland Project (SERP), Somali Regional State of Ethiopia.* Abdi M. Dahir (undated)

Foreign Trees

Exotic trees introduced into the Somali environment

Baxrasaaf	Eucalyptus camaldulensis
Boordi	Cassia siamea
Garanwaa	Prosopis juliflora
Geed hindi	Azadirachta indica
Maacaleesh	Delonix regia
Mahogany, widhwidh	Khaya senegalensis
Mirimiri	Schinus molle
Sabsabaan/Geed Walaayo	Parkinsonia aculeata
Shawri	Casuarina equisetifolia
Xoolo naaxiye	Leucaena leucocephala
??	Albizia lebbeck
??	Sesbania grandiflora

Exotic edible fruit plants introduced

Afakaado	Persea americana Mill
Baashoon	Passiflora edulis
Babaay	Carica papaya
Bambeelmo	Citrus paradise
Baydaan	Terminalia catappa
Biibbo	Anacardium ocddentale
Cambe laf	Mangifera indica
Canbe caad	Annona squamosal
Canbe shoog	Annona muricata
Cinab	Vitis spp
Liin dhanaan	Citrus sinensis
Liin macaan	Citrus sinensis
Liin qarboosh	Citrus aurantifolia
Maandariin	Citrus reticulate
Naarajiin	Cocos nucifera
Rumaan	Punica granatum
Saytuun	Psidium guajava

Edible Wild Food Plants

Plant name	Botanical name	Uses
Afgub	Commiphora tabuk Sprague	Fruits eaten
Canjeel	Mimusops angel Chiov	Fruits eaten
Cillal	Unidentified Asclepiad.	Tubular root eaten
Cismaandoy	Monanthotaxis Formica	Fruits eaten
Carmo gorayo	Cissus sp.	Creeper. It was used as a drought food whereby the leaves where boiled than eatern.
Carrab Lo'aad	Ceropedia sp.?	Branches eatern
Cayo	Gynandropsis gynandra	A small ground creeper; can be eaten in the fresh or cooked in sauce
Berde	Ficus glumosa/ F. Vasata	Fruit eatern
Daray	Ficus sycomorus L.	Fruit eaten
Doombir	Ceropegia sp.	Tuber containing water. It is first skinned and then eaten. Rather bitter
Debnadiir	??	Branches eaten
Dhafaruur	Grewia tenax	Small shrub, the yellow berries of which are eaten
Dhebi dhanaan	Grewia sp.?	Fruit eaten
Dinnaax	?	Root eaten
Gacayro	Caralluna sp.?	Branches eaten
Gocoso	Cyphostemma adenocule	The fleshy root is eaten
Gob	Zizyphus mauritiana Lam.	The yellow and sometimes red cherry-like fruit is

		eatern
Galool	*Acacia bussei*	The swollen bases of the long thorns when green are eatern. These are called Canbuul. Also its flower has sweet nectar.
Garas	*Dobera glabra*	The fruit is boiled and eaten. Takes many hours to cook.
Gommoshe	*Grewia villosa L.*	Berries eaten
Himir	*Ehretia orbicularis*	Berries eatern
Hohob	*Grewia sp.?*	Fruit eaten
Jeerin	*A. edgeworthii*	When green the while pulp inside the seedpod is eatern. When dry the seeds are roasted.
Jinow-dhanaan	*Commiphora rostrata*	Leaves eatern
Kulan	*Balanities spp.*	Fruit eaten. Thejuice is first sucked; this has an agreeable sweet taste but purgative in action. The nut, which composes most of the fruit, is boiled and eaten.
Likke	*Hydnora abyssinica*	This is very peculiar plant that when bursting into flower, forces up the ground (fuur), making a stellar-shapped fracture of the surface, indicating its presence. It is then dug up and eaten
Maddooyaa	*Unidentified*	Berries eatem

		Rhamnaceous plant
Madheedh	*Cordia ovalis/ Corida somalensis*	Fruits eatern
Maydho	*Phoenix reclinata*	Fruits eaten. They are called "Cawaag".
Midhcaanyo	*Grewia erythraea Schweinf.*	Berries eatern
Midha geeljire	*Pavonia arabica*	Fruit eaten
Madheedh		Fruits eatern
Moxog	*Manihot aipi Phol*	Root eaten after cooking
Marooro	*Ceropegia sp.*	Peculiar plant generally found growing out of the centre of a clump of dareemo grass. The bulbous root is eatern as well as the 'trailer."
Murcud (Murcid)[67]	*Ximenia americana L.*	Berries eaten[68]
Ontorro	*Cordyla somalinses*	Fruits eaten
Qoodho orgi	*Canthium Bogosensis (Mart.)*	Fruit eaten
Qudhac	*Acacia tortilis*	Seeds cooked and eaten
Raxanreeb	*Commiphora candidula*	Root eaten
Rooxo	*Coccinea cordifolia Cogn.*	The fruit of the creeper is eaten
Shanfarood	*Garcinia livingstonei*	Fruits eaten
Sholoole	*Uvaria acuminate*	Fruits eaten
Sobkax	*Glossonema*	Fruit eaten

---

[67] *The plant is a shrub said to be localized in Badhan area. (Osman M. Ali)*

[68] *Drake Brochmann (1912) mentioned in his book,* British Somaliland *Mandarug, while Glover (1945) recorded two entries namely,* Mandarut (Chiov.) *and* Mandurud (Chiov.) *in his book,* A Provisional Checklist of Somali Plants.

	boveanum Decne.	
Tiintiin (Tiin)	*Opunita ficus-indica*	Fruits eaten
Timayulukh	*Hydnora spp.*	Tuber; sweet in tase. Ratels are very fond of digging it up and eating it
Timayulukh	*Grewia erythraea Schweinf.*	Fruits eaten
Uneexo	*Cynanchum Somaliense N.E. Br.*	Berries eaten
Unuun	*Citrullus vulgaris Schrad.*	Fruit eaten medicinally, slices are cut and thrown into milk and allowed to soak for some hours, and then the milk is drunk.
Yaaq	*Adansonia digitata*	Fruits eaten
Yicib	*Cordeauxia edulis*	The nuts are very nutritious
Xaadoole	*Cucurbit sp.?*	This plant is eaten raw
Xamakow	*Edithcolea sordida*	Plant eaten like Xaadoole
Xamudh	*Zizyphus hamur Engl.*	Fruit eaten
Xangeeyo	*Sarcostemma viminale*	Branches eatern
Waarig	*Kind of a mushroom*	Flesh eaten. Usually found around anthills. Like the "Likke/Timayulukh" forces its way up the ground. (Waa waarig e , soo weeraraay!"

Names of some fishes in Somali waters

Waqooyi	Bari	Banaadir	J/Hoose	English Names	Scientific Names
Fuluus	Sucbaan	Sucbaan	Fuluus	Dolphin fish	Coryphaena hippurus
Kalbul baxri	Loolaaq	Ey-maayo	Taada	Shark sucker	Echeneis naucrates
Carabi	Carabi	Caanood	Mkidhi	Mullet	Mugilidae spp.
Qud	Faarde	Shooley	Mutumbu	Needle fish	Albennes Hains
Abushook	Cayddi	Cayddi	Dagaa	Sardine	Sardinella fimbriate
Mukhnuus	Maqnaf	Simibilig	Ubabi	Wolf herring	Chirocentrus dorab
Dawaco	Dawaco	Tixsi-gaad	Nuufi	Indian flat head	Platycephalus indicus
Kumal	Gacoorre	Funni	Fumme	Cat fish	Tachysuridae spp.
Caruusa	Fuur	Maambiyo	Boono	Parrot fish	Scarus Ghobban
Siisaan	Ismiir	Saafit	Taasi	Rabbit fish	Siganus spp.
Laba gadhle	Labo-garle	Fangalaati	Imkooma	Goat fish	Mullidae
Xilwa	Xilwa	Xalaawi	Kuuku	Jack pomphret	Formio niger
Sakhlad	Silqo	Taqo ama Silqo	Takaa	Cobia	Rachycentridae
Caqaam	Ganaad	Aluuso/ Subsaalim	Kisumba	Barracuda	Sphyraenidae
Gaxaash	Gaxaash	Cagoole/ Dhuuban	Tangui-jaafa	Emperor Scavenger	Lethrinus nebulosus
Afdheere	Afdheere	Miraamir	Borasimbo	Long face emperor	Lethrinus miniatus
Cawrad	Tunbuur	Shuure-gale	Mkindhi	Mullet	Liza nebulosus
Cursin	Dhocdhocle	Birbirow	Kaawe	Red stripped sea bream	Argyrops filiametosus
Ximaari	Dhocodhocle	Xabkoole	Ki-oofa	Yellow finned sea bream	Acanthopagrus tatus

Diseases of Livestock and burden animals

Sheep and Goats
1. Buraseella
2. Cabeeb
3. Caal
4. Cambaar
5. Candhagooye
6. Boog
7. Darrato
8. Dibbiro
9. Furuq
10. Geed caanoole
11. Sangalle
12. Kuurkuur
13. Raaf-dillaac
14. Oofwareen
15. Sambabfaraq
16. Sunsun (shuban)
17. Hulumbe
18. Qallal (teetano)
19. Qoorgooye
20. Shubanka maqasha
21. Wadne biyood
22. Waallida xoolaha

Cattle
1. Buraseella
2. Cadho
3. Daba-ka-ruub
4. Dibbiro
5. Dhukaan
6. Itaysa (Garbagooye)
7. Raaf-dillaac
8. Qaaxo
9. Qallal (teetano)
10. Cabeed
11. Cadhokar

Camels
1. Afruur
2. Buraseella
3. Cadho
4. Dibbiro
5. Dhugato
6. Furuq
7. Gabdhow
8. Garbabeel
9. Sangalle
10. Kud/Xaaraan/Kaaraan
11. Qaaxo
12. Jajabow/Rigaax
13. Qallal

Horeses and Donkeys
1. Darfac
2. Raaf-dillaac

Selected list of common mammals

Magaca	Magaca cilmiga ah	Magaca Ingiriisiga
Bakayle	Lepus capensis	Rabbit
Balanqo	Cabus ellipsiprymnus	Waterbuck
Bawne	Procavia capensis	Rock Hyrax
Beyrac	Dorcatragus megalotis	Beira
Biciid	Oryx beisa	Oryx
Daayeer	papio hamadryas	Baboon
Calakud	Oreotragus saltatar	Klipspringer
Cawl	Gazella soemmeringii	Soemmering's Gazelle
Yaxaas	Crocodylus niloticus	Crocodile
Dameer farow	Equus gravy	Grevy's Zebra
Deero	Gazella spekei	Speke's gazelle
Deero	Gazella Pelzelni	Pelzelni's gazelle
Dhidar	Crocuta crocuta	Spotted hyena
Dibtaag	Ammodorcas clarkei	Dibtag
Doofaar	Phacochaerus aethiopicus	Wathog
Garanuug	Lithocranius walleri	Garanuug
Geri (hal-geri)	Giraffa camelopardalis reticulata	Giraffe
Goodir/giiryaale goodir (lab), Adeeryo (dheddig)	Strepsiceros koodoo	Greater Kudu
Goodir caarre	Strepsiceros imberbis	Lesser kudu
Gorayo	Struthio camelus	
Gumuburi	Equus nubianus somalicus	Somali Wild Ass
Jere	Hippopotamus amphibious	Hippopotami
Fox	Vulpes pallida	Dawaco

Libaax	Felis leo	Lion
Maroodi	Elephas africanus	Elephant
Sakaaro: Gol cas	Madoqua swaynei	Dik-dik
Sakaaro: Gusuli	Madoqua phillipsi	Dik-dik
Sakaaro: Guyuc	Madoqua guentheri	Dik-dik
Shabeel	Panthera pardus pardus	Leopard
Siig	Bubalis swaynei	Somali Hartebeest
Waraabe	Hyaena striata	Hyaena
Wiyil	Rhinoceros bicornis	Black Rhinoceros

Glossary

Abiotic	Non living
Adaptation, climate change	Actions taken to help communities and ecosystems cope with changing climate condition.
Acid rain	Toxic gases such as sulphur dioxide and nitrous oxide dissolve in rain water and come as acid rain.
Aerobic	An organism that lives and grows in the presence of oxygen.
Afforestation	Establishment of a forest or stand of trees in an area where there was no forest.
Anaerobic	An organism that can live in the absence of oxygen.
Agropastoralism	A way of life of combining growing crops and livestock production.
Acid rain	Toxic gases like sulphur dioxide and nitrous oxide dissolve in rain water to form sulphuric acid and nitric acid and come down as acid rain.
Air pollution	Presence of toxic chemicals in the atmosphere.
Annual plant	a plant that completes its life cycle, from germination to the production of seed, within one year, and then dies.
Aquifer	a layer of permeable rock containing water.
Arid	(of land or climate) too dry to support vegetation.
Badlands	A barren land where erosion has cut loose dry soil or soft soil into strange shapes.
Biodegradable	Substances that can be broken down by bacteria or other living things and thereby causing no pollution.

Biodiversity	The variability among plant and animal species in the world or in a particular habitat.
Biomass energy	Organic matter that has stored energy such as wood, paper, dried vegetation, crop residues, aquatic plants and even garbage.
Biome	A distinct ecosystem characterized by particular plants, animals and climate.
Biotic	Living
Biomedical waste	Waste (be it liquid or solid) generated from biological or medical sources and activities.
Biogas	Any gas retrieved from the decay of organic material.
Biosphere	Zone of earth where organisms live, including the ground and the air.
Briquette	A block of compressed coal /charcoal, or sawdust, leaf litter, burned as fuel.
Bund, soil	A raised structure of earth aimed at preventing soil erosion and conserving water.
Carbon sink	A natural environment, such as a forest, ocean, that have the capacity to absorb carbon dioxide from the atmosphere and store it.
Carrying capacity	In rangeland science, the maximum stocking rate possible.
Check dam	A barrier to retard water runoff for controlling water erosion, and increasing soil moisture.
Climate	The average or prevailing weather conditions of a place over a period of years.
Climate change	A long-term change in the earth's climate, due to increase in average atmospheric temperature.
Climax plant community	The ultimate biotic community formed in an ecological successional series.

Coal	Black or dark brown mineral deposit containing combustible substances that is considered as fossil fuel.
Colony collapse disorder	A phenomenon characterised by the sudden disappearance of a bee colony and leaving behind the queen, resulting in colony death.
Commensalism	A mutualistic relationship whereby one organism benefits from the relationship while the other one is neither benefited nor harmed.
Compaction layer	A near surface layer of compacted soil as a result of repeated impact on or disturbance of the soil surface.
Conservation	System of cultivation that reduces soil and water loss.
Contour farming	A system of farming across slope following the changing slope rather in straight line.
Compost	Nutrient-rich decayed organic matter used as a fertilizer.
Consumerism	The protection or promotion of the interests of consumers.
Contamination	The unwanted pollution of something by another substance.
Coping mechanism	Fall-back mechanisms whereby external and internal stress or threats are managed or adapted to or acted upon.
Coral bleaching	A process by which coral colonies lose colour due to prolonged algae loss caused by warmer water in climate change conditions.
Corporate responsibility	Sense of duty, ethical and transparent dealing of a company toward the environment and social well-being.
Deciduous	Trees that shed their leaves or foliage at the end of the growing season.

Decomposers	Organisms that break down dead plants and animals into inorganic substances.
Deforestation	Clearance or removal of trees or a forest stand.
Degradation	The deterioration of the environment through the destruction and misuse of the natural resources.
Demarcation	The act of establishing a boundary or limits of something.
Desert	A large, barren region, with little precipitation, either with rocky or sandy and little or no vegetation.
Desertification	The persistent degradation of dryland ecosystems that leads to loss of bodies of water, vegetation and wildlife.
Dominant species	Plant species that predominate in an ecological community, in terms of number, coverage, or size.
Drought	A long period of abnormal rainfall leading to water shortage.
Early warning	A major element in disaster risk reduction that provides information on occurring hazards that might involve disasters unless early response is taken.
Ecosystem	A biological community interacting with each other and their physical environment.
Ecotourism	Low-ecological impact tourism that combines tourism with appreciation of nature and wildlife.
Effluents	Liquid waste that is sent out from a factory, farm, commercial establishment, or household into a body of water.
Endangered species	Species considered to be at risk of extinction or has been categorized by the International Union for the Conservation of Nature (IUCN) as Red

	List as likely to become extinct.
Endemism	Restriction of an animal or a plant in a particular area.
Energy flow	One of the ecological processes: The amount of energy that flow through a food chain.
Environmental impact assessment	The process of evaluating the likely impacts of a proposed activity on environment.
Exotic plant	Non-native plant.
Famine	Extreme or general shortage of food.
Family planning	Use of birth control to determine the number of children to be born in a family.
Fauna	All the animals in a particular area.
Fertilizer	A chemical or natural material added to soil to improve its fertility.
Fodder	Animal feed - cut, dried hay for livestock.
Food chain	A series of organisms each depending another for its source of food.
Food security	The state of having reliable access to sufficient, affordable and nutritious food by all people, at all times.
Fossil fuels	Natural fuels formed in the geological past from the fossilization of plants/animals like petroleum, coal, natural gas.
Genetic diversity	The vast numbers of different species, their number and genetic characteristics.
Genetic engineering	Deliberate scientific alteration of the characteristics of an organism, such as adding new DNA into an organism.
Genetically modified foods	Genetically engineered food.

Geothermal energy	Energy generated from tapping underground reservoir of heat.
Global warming	A gradual increase in the average temperature of the atmosphere due to the accumulation of greenhouses gases.
Grazing management	Control of grazing habits on pasture to accomplish a desired purpose.
Greenhouse effect	Trapping of sun's warmth in the lower atmosphere due to greenhouses gases like carbon-dioxide, methane, water vapour etc.
Gully erosion	Removal of soil along drainage creating large channels.
Habitat	An area or the place where an organism lives.
Hardpan	The condition of the subsoil becoming a hard impervious layer.
Herbicides	A chemical used to destroy unwanted plants, especially weeds.
Hydrocarbon	A compound of hydrogen and oxygen.
Hydro-electric power	Electricity generated from hydropower.
Indigenous species	A species whose presence is a region is the result of the natural processes.
Infiltration	Soaking of water into the soil.
Insecticide	A chemical substance to kill insects.
Land degradation	Temporary or permanent decline or loss in the productivity of land through natural and man-made processes.
Landfill	A site for disposal of waste whereby it's buried in layers.
Landslides	Collapse and movement of rocks and soil downhill.
Living fence	A fence made of living trees/shrubs.
Marginal lands	Land of poor quality found on the edge of cultivated or protected lands.
Microclimate	The climate of a small, distinct area differing from the surrounding area.
Migratory birds	Birds that migrate, often north/south

	and vice versa, with seasonal changes.
Mineral cycle (mineral cycle)	The process by which nutrients move from the physical environment into living organisms, and back to the physical environment.
Mist forest	A forest that thrives on the presence of mist.
Mitigation	Action of reducing the severity, frequency and magnitude of a risk.
Natural resources	Materials found in nature that can be exploited for economic gain.
Noxious plant	A plant considered as undesirable and troublesome growing where it is not wanted.
Open grazing (free grazing)	A tract of grazing land without barrier (enclosures/fencing) where livestock and wildlife roam freely.
Over-grazing	To graze in excess of a range's capacity to recover.
Ozone layer	A layer in earth's stratosphere that absorbs most of sun's ultraviolet rays.
Pastoralism	Socio-economic system adapted to dryland where herding is the primary economic activity of a society.
Perennial plant	A plant that can live more than two years.
Photovoltaic	Method of generating electricity from the solar energy.
Palatable species	Agreeable to the palate or taste.
Pollutant	A substance introduced into the environment that pollutes something, especially water and atmosphere.
Precipitation	Any kinds of weather condition where something is falling from the sky on the earth's surface such as rain, snow, ice, etc.
Predator	An organism that naturally feeds on

	another organism.
Prey	An animal that is hunted by another.
Rangelands	Open land for grazing animals and wildlife.
Recycling	Conversion of waste into reusable material.
Reforestation	Renewal of forest cover.
Relict	A remnant of a species that has been widespread in the past.
Renewable energy	Any energy sources that is replenished and not depleted such wind and solar power
Rotational grazing	A type of grazing management where the grazing resource is alternated between grazing and rest periods.
Salt lick	Naturally-occurring salt deposits where animals come go for a lick.
Sand storage dam	A structure built across a dry river bed to increase the volume of sand to store more water.
Semi-arid	It is a climate that is characterized by little annual rainfall.
Shelterbelt	Windbreak. Plants made into rows to reduce wind effect.
Shrub	A woody plant, smaller than a tree that has several stems.
Social responsibility	The obligation of an organization or company to act for the benefit of the society at large.
soil erosion	The wearing and/or blowing away of topsoil.
Soil structure	The way soil particles are arranged in various aggregates such as silt, sand and aggregates.
Solid waste	Unwanted and useless products in the solid state.
Solar energy	Radiant heat and light from the sun.
Species	A group of organisms with similar characteristics that actually or

	potentially interbreed in nature.
Sub surface dam	A structure built build below the surface and across a dry river bed to increase the volume of sand to store more water.
Symbiosis	Association or relationship between two or more living things typically to the advantage of both.
The Tragedy of the Commons	A problem whereby every individual tries to get the most out of a common resource thus leading to its early degradation.
Terraces	Flat areas built on a slope, hill or mountain for the purpose of checking water runoff and for cultivation.
Tillage	Agricultural reparation of land.
Toxic waste	Poisonous waste material.
Transhumance	Seasonal movement of people with their livestock herds from one place to another.
Ungulate	Hoofed herbivore.
Unpalatable species	Not please to taste.
Urbanization	Increasing concentration of people in urban areas.
Vegetation zone	A band of land having physiognomically similar vegetation type.
Waste disposal	Proper handling of discarded waste in a way it does not cause harm to the health and environment.
Water cycle	The process by which water is circulated throughout the earth.
Watershed	The entire area draining under gravity into low lying areas - seasonal water

courses and rivers in particular.

Emergency Water trucking	Transporting water on trucks for life-saving operations.
Weather	Description of the state of the atmosphere in a particular area.
Weed	A valueless plant growing wild and limiting the growth of useful plants.
Wetlands	Areas with standing water covering the soil and having rooted vegetation.
Wild foods	Edible plants in the wild.
Wildlife	Undomesticated animals in the wild.
Wind-break	Shelterbelt.
Wind energy	Electrical energy harnessed from the wind.

Bibliography

1. **Abdi M. Dahir,** *Herbarium Collections at South Eastern Rangeland Project (SERP),* Somali Regional State, Ethiopia.(undated report)
2. **Academy for Peace and Development (APD),** *Regulating the Livestock Economy in Somaliland,* Hargeysa 2002
3. **Alisha Ryu,** Waste Dumping off Somali Coast May Have Links with Mafia, Somali War Lords.
4. **Annarita Puglielli** iyo **Cabdalla Cumar Mansuur,** *Qaamuuska Afsoomaaliga,* Centro Studi Somali, Università degli Studi Roma Tre, *RomaTrE-PRESS 2012*
5. **Awale, Ahmed Ibrahim & A. J. Sugulle,** *Perennial Plant Mortality in the Guban Areas of Somaliland,* Candlelight study 2011.
6. **Awale, Ahmed Ibrahim,** *Climate Change Stole Our Mist.* Candlelight NGO, Hargeysa, Somaliland (2007)
7. **Awale, Ahmed Ibrahim et. al:** *Impact of Civil War on the Natural Resources: A Case Study for Somaliland (2006),* Candlelight NGO, Hargeysa, Somaliland.
8. **Awale, Ahmed Ibrahim & Sugulle, Ahmed J.:** *Invasion of Prosopis juliflora in Somaliland: Challenges and Opportunities. Candlelight NGO, Hargeysa, Somaliland.(2006)*
9. **Awale, Ahmed Ibrahim** et. al: Case Study: Integrated Community-based Resource Management in the Grazing lands of Gacan Libaax, Somaliland.
10. **Bally, P.R.O., & Melville, R.** Report on the Vegetation of the Somali Democratic Republic with Recommendations for its Restoration and Conservation, Dec. 1972

11. **Couter, J,** 1987, *Market Study for Frankincense and Myrrh from Somalia,* Tropical Research and Development Institute (TRDI)
12. **Duale, Omer H. & Magan, Abdillahi H**. : *Case Study: Alternative Source of Energy and Reduction of Dependence on Charcoal in Somaliland,* Candlelight Hargeysa, Dec. 2005
13. **Gilliland, H.B.,** *The Vegetation of the Eastern British Somaliland. The Journal of Ecology,* Vol. 40, No. 1, (Feb., 1952), pp. 91-124.
14. *Glover*, P. E., – 1947 – A Provisional *Checklist* of British and Italian *Somaliland* Trees, Shrubs and Herbs. Crown Agents, for the Government of *Somaliland*
15. **Ingrid Hartmann et. al.,** *The Impact of Climate Change on Pastoral Communities in Salahley and Balli Gubadle Districts. Candlelight study (2011).*
16. **Hemming, C. F.,** *The Vegetation of the Northern Region of the Somali Republic. Anti-Locust Research Centre, London. January, 1966*
17. **Herzog, M.,** *Forestry and Woodland Management in Somaliland: Problems, Background, Developmental Potentials,* Caritas Switzerland, Lucerne, 1996
18. **Karekezi S. & Ranja T.,** *Renewable Energy Technologies in Africa,* AFREPREN, Zed Books, (1997)
19. **Klughardt, Doris and Killeh, Mohamed E**. *Community Based Rehabilitation of Wadi Management in Baki district,* Awdal Region, German Agro Action Somaliland, 2002
20. **Lemma Belay et. al.,** *The Impact of Climate Change and Adoption of Strategic Coping Mechanisms by Agro-Pastoralists in Gabiley Region, Somaliland.* (Candlelight study 2011)
21. **Leslie A**. D., *An Introduction to the Woody Vegetation of Somalia;* British Forestry Project Somalia Research Section, Working Paper 11, (1990), (NRA/ODA)

22. **M. Roderick Bowen**, *a Survey of Tree Planting in Somalia – 1925-1985,* (1988). Oxford Forestry Institute Occassional Papers, University of Oxford
23. **Malte Sommerlatte and Abdi Umar**, *An Ecological Assessment of the Coastal Plains of North Western Somalia (Somaliland)*, IUCN, 2000
24. **McCarthy, Gerry et. al**, *Somali Private Sector Appraisal and Recommendations,* May 2005, International Finance Cooperation & the World Bank
25. **Miskell, John.** *An Ecological and Resource Utilization Assessment of Gacan Libaah, Somaliland,* IUCN Eastern Africa Programme, Nairobi (May 2000)
26. **Mohammed Ibrahim Abdi,** *Critical Health and Environmental Issues Relating Leather Industry in Somaliland, Preliminary Environmental Assessment Report* (June, 2014)
27. **Muna Ismail**, **Lewis Wallis** and **Scot Draby**, *Restoring Land and Lives: Report of a scoping mission to examine the restoration and possible domestication of the Yeheb plant in Somaliland,*(June 2015), Initiatives of Change, London, UK.
28. **Sadia M. Ahmed**, *Survey on the State of Pastoralism in Somaliland, PENHA/ICD, 2001*
29. **Thulin, Mat**, The Flora of Somalia, 4 volumes. Kew Royal Botanic Gardens (1993).
30. **Wixon, Calvin,** *Arabsiyo Soil and Water Conservation Project,* (Somali Republic), USAID Dec. 1964

Other documents

1. Cimilo-Awaal: Daraasado ku saabsan Doorsoonka Cimilada, Candlelight 2011

2. Country Economic Report for Republic of Somalia, Synthesis Draft, World Bank June 30, 2005
3. *DEEGAANKEENNA (Our Environment) Newsletter*, Issues 1-41, Published by Candlelight Org.
4. *Envirnomental Assessment on the Gebi Valley and the Sool Plateau Sanaag region (Somaliland-Puntland)*, Horn Relief study (2005).
5. Land Tenure Policy Workshop, VETAID Somaliland (July 1997)
6. Land Resources Tenure and Agricultural land use MoPDE & MoA August 2002.
7. *Natural Resources Protection and Conservation Act No. 04/98*, Ministry of Pastoral Development and Environment, Somaliland.
8. Northwest Agricultural Development Project, Feasibility Study and Technical Assistance. Technical Report No. 6 Range & Livestok Study, SOGREAH, June 1982.
9. *Report of the Panel of Experts in Somalia, Pursuant to Security Council Resolution 1474 (2003), PP. 59*
10. Somalia: *Towards a Livestock Sector Strategy*. FAO, World Bank, EU Mission in Kenya, Report No. 04/001 IC-SOM, 2004
11. Somalia Agricultural Sector Survey, Main Report, World Bank, December 1987.
12. Somali Democratic Republic, Land Tenure Law No. 73
13. Somaliland Land Ownership Law 08/99.
14. Somaliland Environmental Policy (2012)
15. Somaliland Range Policy (2001)
16. The Laws of Somaliland Protectorate, Chapter 119, Cultivation & Use of Land, Jan. 1950
17. Traditional Food Plants of Kenya (National Museum of Kenya, 1999, 288 p.)

18. Yearly Fisheries & Marine Transport Report, 1987/88, Ministry of Fisheries & Marine Transport, Somalia Republic.

Index

Abdullahi Ashur, 35
Acacia tortilis ('*Qudhac*'), 64
Action in Semi-arid Lands (ASAL), 134
Adaptation, climate change, 60
Agave sisalana, 72
Al Wabra Wildlife Preservation (AWWP), 62, 63
Al-Ain, 134
Aloe eminens (*Dacardhaadheer*), 26, 27
Alpha-HCH, 22
Antarctic, 58
Aqal (Somali hut), 78
Arid and semi-arid lands (ASAL), 59
Asal, 37
Ayaana, spirit of, 91
Ayaha Valley, 21
Baboons (*Papio hamadryas*), 149
Balanites orbicularis ('*Kulan*'), 64
Bangladesh, 58
Baraawe, 54
Barack Obama, President, 122
Barzakh (Purgatory), 154
Basra, port of, 131
Bees, 18, 85, 86, 107, 108, 128, 143, 144
Beira antelope, 62
Berbera, 29, 41, 42, 64, 65, 72, 82, 88, 93, 94, 116
Beta-Endosulfan, 22
Beta-HCH, 22
Biodiversity loss, 55
Biomass, 53
Birds, pesticide contamination of, 123
Biyofadhiisinka, 72
Boorama, 59
Boosaaso, 54, 130, 132, 133, 134, 136
Boungavillea, 68
British Broadcasting Coorporation (BBC), 63
Brochmann, Drake, 150
Bullaxaar (Bulhar), 45, 65
Burco, 59
Butzer, K. W., 74
Buxus hildebrandtii ('*Dhosoq*'), 27
Cal Madow, 71
Cancer, 18, 23, 120
Candlelight, 5, 12, 44, 46, 80, 90, 138, 155, 201, 202, 203, 204
Carbon sequesteration, 55
carbon sink, 33, 55
Caynabo, 63
Ceelaayo, 54
Ceel-macaan, 54
Ceerigaabo (Erigavo), 26
charcoal, 17, 28, 29, 33, 47, 49, 50, 53, 54, 55, 73, 81, 82, 83, 84, 87, 91, 104, 110, 143, 144, 152
Charcoal production, 49, 53, 82, 104
Christ's Crown of Thorns, 18
Civil society organizations (CSOs), 54
Climate Change, 56, 58, 65, 80, 201, 202
Colony collapse disorder, 85
Common Fisheries Policy of the EU countries, 101
Coral bleaching, 58
Cordeauxia edulis ('*Ye-eb*'), 96
Cordeauxiaquinon, 97
Cushitic belief, 91
Dacar-Budhuq, 93
Daedalus and Icarus, 124
Date palm, 130, 132
Date palm cultivation in Puntland, 131
DDT, 22, 127, 128
deforestation, 36, 42, 50, 51, 53, 84, 86, 104, 105, 135
Deforestation, 49, 50, 52, 103
Dengue, 59
Deri Maraa, 72

Desert Locust Control Organization for East Africa (DLCO-EA), 21
Deyr', 75
Dhuusa-ma-reeb, 98
Dieldrin, 22
Djibouti, 30
Draceana schizantha (Mooli), 71
Drought, definition of, 104
Durduri, 131
Ecotourism, 79
Eden, the Garden of, 153
Effluent treatment, 93
Eldrin, 22
Environment, definition of, 119
Environmental Compliance, 61
Environmental Impact Assessment, 95
Environmental liability, 61
Environmentalism, definition of, 119
Eritrea, 30
Ethiopia, 65, 68, 96, 98, 201
Euphorbia grandis ('Xasaadin'), 71, 79
Extreme weather conditions, 58, 60
Fig tree (*Ficus vasta*), 145
Flooding, 51
Forestry and Wildlife Act, 83
Frankincense, 28
Gaarriye, Mohamed Hashi Dhamac, 154
Gacan Libaax, 60, 66, 70, 72, 74, 75, 76, 77, 79, 82, 83, 84, 201
Gallaaddi (Galadi), 154
Gebilay, 69
Global warming, 39, 55, 56, 58, 60, 105, 120
Glover, P. E., 76
Go'da Weyn, 72
Go'da Yar, 72
Gob (Zizyphus mauritiana), 17, 18, 34

Golis Range, 50, 66, 136, 138
Greenhouse gases (GHGs), 55, 56, 57, 58, 60, 120
Gu' (Spring season), 75
Guano, 117
Guban, 28, 60, 64, 135, 146, 201
Gudmo Biyo-Cas, 31
Gulley erosion, 55, 105
Haji Aadan Af-qallooc, 30
Halin (Xalin), 133
Hargeysa, 12, 13, 14, 15, 21, 22, 23, 24, 25, 29, 33, 34, 35, 36, 37, 38, 39, 46, 59, 68, 82, 88, 90, 92, 93, 112, 145, 155, 201, 202
Hartebeest, Swayne's, 62
Heinrich Boell Foundation, 80
Hemming, C. F., 70, 74, 202
Henna domestication, 137
Henna, medicinal uses of, 137
Heptachlor, 22
Higlo (*Cadaba heterotricha*), 147
Honey granulation, 109
Honey testing, 107
Honey, chloramphenicol-tainted, 107
Horn of Africa, 17, 27, 29, 62, 104, 146
Hunt, John A., 73, 74
Hydrological cycle, 51
Illegal, unreported and unregulated fishing (IUUs), 101
Indigenous knowledge, 79
Infra-red radiation, 56
Inter-governmental Panel on Climate Change, 58
Iraq, 131
Iskudar, 72
Ismail Mire, the poet, 99
Jama, Abdi Ali, 115
Jameeca Weyn, 35
Jiilaal (winter season), 34, 73, 74, 75, 78

Juniperus procera ('Dayib'), 27, 29, 60, 66, 71, 73, 78, 79
Kenya Plant Health Inspectorate Services (KEPHIS), 21
Kerosene, 55
Kidi tree (*Balanities aegyptica*), 152
Kismaayo, 54
Laas Geel, 93
Laas Qoray, 131
Landfills, 58
Landslides, 51
Land-use change, 58
Lindane, 22
Liquefied petroleum gas (LPG), 55
Lord Delemare, 35
Lucy Larcom, 67
Mahatma Gandhi, 64, 149
Malaria, 59
Man, God's vicegerent on earth, 11, 119
Maraa (Acacia nilotica), 14
Marka, 54
Maydh (Mait), 6, 28, 115, 116, 117, 118
Methane, 58
Miskell, John, 70, 203
Mitigation, climate change, 60
Mohamed Ibrahim Egal, 63
Muhammad, the Prophet, 67, 110, 153
Muse Aw Ahmed, 77
National Environment Act, 83
Nayruus (Norouz), 80
Netherlands, 58
Noise pollution, 155
Oil slicks, 120
Oman, 44, 126, 131
Oodweyne, 19, 63
Organochlorine pesticides, 22
Oryx, 103, 189
Ozone layer, 59, 120
Peace and Development Organization, 54
Persia, 110

Photosynthesis, 38, 57, 120, 145
Puntland, 6, 32, 54, 131, 132, 133, 136, 204
Qat (Catha edulis), 15, 28, 49, 72, 77, 104, 155
Qatar, 62
Queen Hatshepsut, 146
Raas Caseyr, 27
Rachel Carson's Silent Spring, 128
Ramsar Convention, 118
Rayne, H., Major, 34
Rural-urban migration, 80
Sahil Region, 50
Salvadora Persica ('*Caday/Rumay*'), 110
Selective fishing practices, 101
Sha'ab village, 32
Sheel erosion, 55
Sheikh Madar, 36
Sheikh Saoud bin Mohammed bin Ali Al-Thani, 62
Sheikh town, 36, 49, 50, 52, 62, 146
Sitaad, 90, 147
Socotra, Island of, 131
soil erosion, 51
Solar cookers, 55
Solid Waste Act, 83
Somali Language Committee, 154
Somali Relief Society (SORSO), 133
Somali Wild Ass, 30, 31, 32, 62, 189
Somalia, 26, 46, 54, 55, 65, 68, 96, 97, 98, 100, 102, 131, 133, 138, 202, 203, 204, 205
Somaliland, 12, 21, 22, 23, 24, 26, 27, 29, 32, 34, 35, 44, 45, 62, 63, 70, 74, 79, 80, 82, 83, 92, 93, 115, 116, 118, 126, 127, 137, 138, 150, 154, 155, 180, 201, 202, 203, 204
Somaliland Environment policy, 83
Sool Region, 32

South/Central Somalia, 54
Soviet Union, 65
Surad Mountain, 26
Swayne, H.G.C., 34
Tanneries, 92
termites, ecological benefits of, 149
Thorn enclosures, 82
Toxic dumping, 102
Traditional weather forecasting, 79
trampling (foot effect), 149
Ultraviolet radiation, 59
United Arab Emirates (UAE), 134
United Nations Development Programme (UNDP), 22

Usnea articulata (Lichens), 73
warthog (*Phacochoerus aethiopicus*), 149
Watershed, 51
Wetlands International, 116
World Conservation Union (IUCN), 30
World Health Organization (WHO), 23
Xagaa (Summer), 75
Xalwo Cige, 70, 77
Xiddigiye, 80
Ye-eb, domestication of, 98
Yellow fever, 59
Yemen, 97, 131, 137, 138
Zeila ('Saylac'), 125
Zizyphus honey, 86

Other books by the author

1. *Environment in Crisis: Selected Essays on Somali Environment/Qaylodhaan Deegaan: Qoraallo Xulasho ah.* PonteinvisibleRedsea-Online.com, Pisa, Italay ISBN #: 88-88934-13-8
2. *Dirkii Sacmaallada* (2012): *Meel-ka-soo-jeedka Soomaalidii Hore: Sooyaal, Rumayn, Ilbaxnimo.* Liibaan Publishers, Denmark. *ISBN #:* 978-87-995208-1-7
3. The Mystery of the Land of Punt Unravelled, Liibaan Publishers, Denmark. ISBN # : 978-8799520848
4. *SITAAD: Is-dareen-gelinta Diineed ee Dumarka Soomaaliyeed (2013),* Liibaan Publishers, Denmark ISBN #: 978-87-995208-2-4
5. *Maqaddinkii Xeebaha Berri-Soomaali (2014),* Liibaan Publishers, Denmark ISBN #: 978-87995208-3-1
6. Qaylodhaan Deegaan: Qoraalo Xul ah, Liibaan Publishers, Denmark, ISBN: 978-87-995208-6-2

Made in the USA
Columbia, SC
20 December 2024